Lipids in Evolution

MONOGRAPHS IN LIPID RESEARCH

David Kritchevsky, Series Editor
Wistar Institute
Philadelphia, Pennsylvania

FUNGAL LIPID BIOCHEMISTRY
By John D. Weete

LIPID METABOLISM IN MAMMALS, Volume 1
Edited by Fred Snyder

LIPID METABOLISM IN MAMMALS, Volume 2
Edited by Fred Snyder

LIPIDS IN EVOLUTION
By William R. Nes and W. David Nes

A Continuation Order Plan is available for this series. A continuation order will bring delivery of each new volume immediately upon publication. Volumes are billed only upon actual shipment. For further information please contact the publisher.

Lipids in Evolution

William R. Nes

Drexel University
Philadelphia, Pennsylvania

and

W. David Nes

Western Regional Research Center
U. S. Department of Agriculture
Berkeley, California

PLENUM PRESS • NEW YORK AND LONDON

Library of Congress Cataloging in Publication Data

Nes, William R
 Lipids in evolution.

 (Monographs in lipid research)
 Bibliography: p.
 Includes index.
 1. Lipids. 2. Lipid synthesis. 3. Microbial lipids. 4. Evolution. I. Nes, W. David, joint
author. II. Title. [DNLM: 1. Evolution. 2. Lipids. QU85 N457L]
QP751.N47   574.1'9247   79-25104
ISBN 0-306-40393-5

QP
751
.N47

© 1980 Plenum Press, New York
A Division of Plenum Publishing Corporation
227 West 17th Street, New York, N.Y. 10011

Printed in the United States of America

To
Estelle and Sandra

Preface

A turning point seems to have been reached recently in our understanding of biological systems. After about 1930, when it became possible to examine the dynamic aspects of a cell seriously and to increase the breadth of our knowledge of the chemicals which are involved, there was a feeling that the elucidation of a pathway or the use to which a compound or a process was put biologically had a ubiquitous character. Among the reasons for believing this was the constancy of the amino acid building blocks for proteins. Not only were the same general structures found regardless of organismic type, but the stereochemistry was the same. This sort of observation led to the idea of the "unity of nature." Few people, of course, thought a complete unity existed, because it was already known in the latter part of the nineteenth century that the sterols, the polysaccharides, and the pigments of various organisms could be substantially different, but only recently, during the last decade or so, has the full scope of the difference—as well as of the similarity—begun to emerge. Of particular importance, it has now become evident that a large hiatus exists between some types of organism. Even more important, perhaps, and certainly more unexpected, are the substantial gaps which have been discovered within otherwise similar organisms. The evolutionary process is presumably behind what we observe, and therefore an understanding of the rules and regulations which govern similarity and difference should eventually lead to an elucidation of evolution itself.

The purpose of this book is to examine the extent to which lipids are or are not the same throughout nature and, where possible, to correlate the information with evolutionary thought. Our review is an outgrowth of a course one of us has been teaching at Drexel University and began as a printed subject in response to an invitation from Dr. David Kritchevsky to write a chapter for *Advances in Lipid Research*. In preparing the manuscript, we found the mass of material too large to bring together in the chapter format. Dr. Kritchevsky was kind enough to say that our efforts ought not to be condensed and suggested we publish instead in the form of this volume in the series *Monographs in Lipid Research*, of which he is also the editor. We are most grateful for this opportunity.

There appear to be several factors which determine the biochemical character of a cell. They include time, lineage, function, and environment. It is not fully clear how separate each of these is. The last two are probably interrelated in a particularly intimate way, but there can be no doubt that the first is paramount. For this reason we have begun our discussion with a review of the evidence for a scale of time and for the documented changes which have occurred at various periods. We proceed next to an examination of primitive character and then on to the lipids of various kinds of organisms. A critical analysis of the significance of the information to evolutionary parameters is attempted throughout. However, the path to an understanding of relationships proves to abound with pitfalls, and in the section dealing with conclusions we have been rather circumspect. The problem can be illustrated with cholesterol. Thought originally to be the zoosterol, it is now found as the dominant sterol of some algae and to be incapable of supporting the life of some insects. Clearly, the genetics for cholesterol biosynthesis and function do not easily fall into a pattern, and a "conclusion" is avoided. We have instead tried primarily to point out the facts in the body of the text along with what consistencies and inconsistencies exist in the hope that the reader will be drawn by the known into the fascinating realm of the unknown, where the final answers lie.

This book would not have been possible without an active research program, and the financial support of the National Institutes of Health is gratefully acknowledged. Thanks are also due Mrs. Velma Moultrie for her devoted efforts in typing the manuscript.

William R. Nes
W. David Nes

Philadelphia and Berkeley

Contents

CHAPTER 1

Introduction

Evolution, for reasons that are not entirely obvious, has once more become a fashionable subject. Certainly a contributing factor is the wealth of data that has become available recently, enough in fact to impinge on our conceptual understanding of the subject. A great deal of attention has been given to information derived from protein sequencing (Dickerson and Geis, 1969) and polynucleotide structure (Darnell, 1979; Hori and Osawa, 1979). It also happens that much is now known in the lipid field, and the purpose of our review is to summarize the contributions made with this class of biochemical.

Charles Darwin (1809–1882) is usually associated with the beginnings of evolutionary thinking. His contributions, however, would have been without consequence had it not been for the intense interest in the topic which had already been generated by his predecessors. While this is not the place to review the chronology, it is important to recall that evolution has had a long history and that serious considerations of it predate our current view of chemistry, to which, along with our current view of physics, we give our greatest allegiance these days. This allegiance is hardly misplaced, since an elucidation of evolution can certainly not be achieved without a clarification of the mystery of matter in both a structural and an energetic connotation. A third factor of major importance is time, or more precisely, the scale of time. In chemistry and physics time is measured in units short enough to be experienced by the experimenter, but evolutionary time staggers the most robust imagination. Yet, a given human being must place the billions of years since life's origin in his particular frame of reference, even though his own life-span is 7 orders of magnitude less than that. Had it not been for the Conte di Quarenga Amedeo Avogadro (1776–1856), who led us to entertain the idea of no less than 23 orders of magnitude, it may be that nothing of consequence in evolution could have ensued. Luckily for those of us alive today, we are a century and a half beyond Avagadro, and we can take it without a flinch when we are told that even our own hominid evolution has taken as long as a million years.

Time. This is the essence of evolution. If we ask where we came from or whether we have come, indeed, from anywhere at all, the question is meaningless without a concept of time, then a concept of change in time, and finally, a view of goals, if such exist, achieved in time. Ernst Mayr, writing recently in a popular medium (Mayr, 1978), has said that "evolution implies change with continuity, usually with a directional component." We, as he, are not prepared to present any more of an all-inclusive definition than this. The idea of evolution has only implications, albeit transcendent ones, placed in a time matrix that can be dealt with only by an examination of what is, together with what has been.

Our review, therefore, really begins with documentation of the scale of time and the changes that have occurred. The last few years have witnessed profound achievements in dating. In parallel, a foundation has been laid recently for the occurrence of properly dated molecular artifacts of previous forms of life as well as for the existence of life itself early in the Precambrian. It is essential to know beyond doubt that life has existed for some particular length of time and that change has occurred, before any other consideration is given to evolution. Consequently, we next review the paleontological record. This is followed by a consideration of the present. Here, much more data is available, because the most remarkable of all aspects of biology is that there remain extant today representatives of all of the major types of living systems that ever lived. We are, in fact, immersed in an evolutionary garden on this planet from which the apples can be picked with profit so long as we learn and remain wary of the rules.

For the serious student of evolutionary thought, a recent book on fossils (Rudwick, 1972) is recommended as an introduction to the various climates of opinion in which evolution has been cast at different periods. It is most instructive to be reminded, as pointed out by Rudwick (pp. 156–157), that by about 1830, which is 150 years ago and at the time when Darwin's father was trying to discourage him from making his now famous trip around the world on the Beagle, the following intellectual barriers had been crossed:

1. Morphology and classification of plants and animals was achieved at a sophisticated level.
2. Fossils had long been discovered and well characterized.
3. Stratigraphy was clearly recognized as time-dependent and hence seen as a chronologic measure.
4. The scale of time was established as far exceeding human life spans, though precise numbers were not yet in hand. Darwin himself actually

thought the Wealden Valley in southeastern England might be as much as 300 million years old.

5. The same general kinds of forms of life were recognized around the world.
6. Changes were perceived as occurring in time with regard to the kinds of living forms.
7. Biological changes could be associated with geological changes.
8. Earlier forms of life were perceived as less advanced than those which came later; in particular, a progression of processes leading from fish to amphibians to reptiles to marsupials to placental mammals and man was apparent along with the succession from cryptogams to dicotyledons in the plant kingdom.
9. The earth itself was understood as having undergone physical evolution, probably from a hot to a cool body.
10. Selective breeding as a means for manipulating inheritable characteristics had long been established.

This reservoir of conceptual information was stymied, however, by the paucity of our knowledge of chemistry. It was only in this same period (1828, to be exact) that Wöhler first proved living systems actually are constituted of chemicals that could be constructed from nonliving matter, and it was two generations away from Pregl, who at the end of the 19th century devised microanalysis as a quantitative means for determining (just) the elemental constitution of biochemicals. Not in fact until 100 years after 1830 did the work of Pregl and a host of others allow the structure of, say, a sterol to be determined.

Structural information is at the heart of modern biochemistry. It reached its flower in the 1950s (see, for instance, Fieser and Fieser, 1959) and permitted an exacting analysis of biosynthesis and other metabolism in the third quarter of the present century (see, for instance, Nes and McKean, 1977). Among the many things which have rapidly emerged from this is the discovery that biochemically all living systems have a unity in general but not necessarily in particular. The differences that occur are associated with phylogeny and with function, and function is associated with environment as well as with time. Of great importance has been the realization that ecological pressures ought to affect only the functional aspects of an organism. The manner in which a given function is arranged in response to environmental factors is conceptually another matter and appears to be a most cogent marker of the evolutionary line by which an organism has ascended the chronological ladder. Lipids have offered an unusual opportunity to examine these problems, since their structures and biosynthesis are now more or less completely understood and since it is they, unlike genetic material and enzymes, which actually perform the functions accruing to the benefit of the organism possessing them.

CHAPTER 2

The Nature of Lipids

The lipid class was conceived long ago as a separate group of biochemicals that remained in the organic layer when macerated living material was partitioned between water and an organic solvent such as ether, benzene, or chloroform. This led to a simple operational definition: lipids are more soluble in organic than in aqueous media. A more important theoretical definition has emanated from these experimental parameters. Since solubility depends on the amount and kind of polar functional groups, lipids must have relatively little capacity for hydrogen bonding or dipole–dipole interactions and a large capacity for van der Waals' interaction. Consequently, lipids are now recognized simply as molecules in which the proportion of C atoms that possess no polar functionality (OH, NH_2, etc.) is large regardless of the question of solubility. The molecules may be acyclic, linear, branched, mono- or polycyclic, and what polar functions there are may be quite varied. The great interest which has developed in these substances comes from the realization that the low degree of polarity inherent in their classification is the key to their biological function. Furthermore, it has become apparent that the nonpolar parts of the molecules, i.e., the hydrophobic moieties, frequently exert their biological significance in a nonmetabolic way, as in the membranous role of fatty acids and sterols. This is in sharp contrast to the previous trend of thought in biochemistry that was focused on molecular changes associated with processes such as energy production or biosynthesis.

Typical lipids are the linear hydrocarbons, their oxygenated derivatives, e.g., fatty acids and alcohols, the sterols, and the triterpenoids and other isopentenoids. In many cases these substances occur in union with a more polar substance as with the phospholipids, sulfolipids, and steroidal glyosides. Our attention in this review will be focused on the less polar moieties of these molecules. We have attempted to use as simple nomenclature as possible in discussing chemical structures, but for the more complex rules used the reader is referred to the official nomenclature of the International Union of Biochemistry (Liebecq, 1978), and also in the case of sterols to Nes (1977), Nes and McKean (1977), and Fieser and Fieser (1959).

CHAPTER 3

Dating and Chronology

A. ORIENTING REMARKS

In 450 B.C., Herodotus realized that the sediments left by the annual flooding of the Nile were responsible for the deposits on either side of the river and that the depth of the sediments indicated a probable age of thousands of years. A numerical value was not forthcoming, however, for another two millennia, when in 1854 the foundation of the statue of Ramses II at Memphis was found to be buried under 9 feet of sediment. Historical records proved the statue to be 3000 years old, leading to a rate of accumulation of 3×10^{-3} feet per year. The upper alluvial deposit covering other strata at Memphis was 40 feet deep, from which it was apparent that, if the rate were constant, some 13,000 years would have elapsed during the formation of this upper layer. This ability to derive numbers marked a great advance in our understanding of the scale of time. In the previous decades it had been appreciated from European mining and from other observations in Germany, France, and Switzerland that the earth was often layered. Since rivers cut through these layers while mountains clearly had carried them upwards, the layers must predate the rivers and mountains. The further deduction was made that the layers represented successive periods of time in a scale much larger than human life-spans. Dating, then, by alluvial deposits offered a method of semiquantitation and placed geologic time in the tens of thousands of years at least. In the 1860s the scale was further extended by Lord Kelvin. Based on heat flow from the center of the earth, he estimated geological age as no less than 25–40 million years old. Not knowing exact values for density gradients, etc., he could not have been expected to have made an exact calculation, and his number was still further expanded in 1899 by estimates of the age of the ocean based on the concentration of salt in rivers and the annual rate of flow. It was concluded

Table 1. Some Radioactive Elements Used in Dating

Element	Half-life (years)	Decay process	End product
Rubidium-87	47×10^9	β Emission	Strontium-87
Rhenium-187	44×10^9	β Emission	Osium-87
Thorium-232	14×10^9	Long series	Lead-208
Uranium-238	4.5×10^9	Long series[a]	Lead-206
Potassium-40	1.3×10^9	Electron capture	Argon-40
Uranium-235	0.7×10^9	Long series	Lead-207
Carbon-14	5.7×10^3	β Emission	Nitrogen-14

[a]The first step (α-particle emission) in the decay series of $^{238}U_{92}$ is the formation of $^4He_2$ and $^{234}Th_{90}$. Several subsequent steps also emit helium.

that the ocean is 100 million years old. While the exact reliability of this number also suffered from inaccuracies, notably in the estimate of the total volume of river flow, constancy of the rate, and the effect of dry seas, the size itself of the number, 100 million years, placed our view of the history of the earth in still a new dimension and one which is only a factor of ten less than the value currently accepted.

Stratigraphy (the study of succession as seen in layered geologic formations), beginning seriously with the work of William Smith (1769–1839), had been established for about 100 years by the time the estimate was made of the age of the ocean (giving also a minimal age for the earth). What was needed was a method for dating the strata themselves as well as methodology to assure constant rates. Both were soon forthcoming. In the present century, through the use of isotope ratios, especially those involving radioactivity (Table 1)—dating methods have been brought to highly sophisticated and varied levels. As outlined in the rest of the chapter; it has been possible to date the whole sweep of evolutionary history (Table 2).

B. THE AGE OF THE UNIVERSE

The universe consists of perhaps 30 billion galaxies, two-thirds of which are flat discs with two spiraling arms. In 1924 Edwin P. Hubble (1889–1953) first showed the existence of a galaxy (Andromeda) beyond our own (the Milky Way), the separation factor being about ten times the cross section of the Milky Way. His paper was presented to the American Astronomical Society on December 30, 1924, and was abstracted the following year (Hubble, 1925). Later he and others showed that within about 800 megaparsecs from us (15% of the universe) there are some 20 million average galaxies. (One parsec equals 3.26 light years, or 3.09×10^{18} cm.) The galaxies exist in a continuous hierarchy from individual ones through groups of

Table 2. Geologic Time Designations

	Eras	Periods	Epochs	Age of oldest part of period (in 10^6 years)
Phanerozoic	Cenozoic	Quaternary	Recent Pleistocene	0.025 1–2
		Tertiary	Pliocene Miocene Oligocene Eocene Paleocene	11–13 25 36–40 58–60 63–70
	Mesozoic	Cretaceous	Upper Lower	110 135
		Jurassic	Upper Lower	166 180–181
		Triassic	Upper Lower	200 225–230
	Paleozoic	Permian	Upper Lower	260 270–280
		Carboniferous Pennsylvanian Mississippian		 310 345–350
		Devonian	Upper/Middle Lower	390 400–405
		Silurian		425
		Ordovician	Upper Middle/Lower	445 500
		Cambrian	Upper Middle/Lower	530 600
	Archean	Precambrian		3300

galaxies to clusters of groups of galaxies. (For a discussion of galactic distribution, the properties of galaxies, and the significance of galactic clustering to cosmology, and for a key to the literature, see Geller, 1978, and Larson, 1977. For a discussion of isotopic abundances within our own galaxy and their relation to the record of the past, see Bertojo *et al.* (1974). For a review of interstellar hydrogen and its impact on our knowledge of galaxies, see Roberts, 1974.)

Three main views of the history of the universe have developed from modern science. They are (1) a "steady state" universe, propounded especially by the English astrophysicist Fred Hoyle, in which destruction of matter is counterbalanced by creation of matter, in a universe with neither a beginning nor an end; (2) an evolving and expanding universe with a beginning, in a "big bang," and no apparent end; and (3) a universe beginning, ending, and repetitively evolving through new beginnings and endings (the pulsating universe). While no way to assess the recycling or third possibility is in hand, evidence that the universe actually had a beginning developed from discovery of the famous "red shift" in the wavelengths of light reaching us from outer space. This was interpreted as a Doppler effect (see Roberts, 1974, for a recent discussion), which meant the volume of the universe must be expanding as a result of the pieces of matter constituting it moving away from each other. By extrapolation from the observable rate of expansion, such a view suggested that at some prior time the process had begun. Hoyle's alternative interpretation of the red shift (an interpretation he now feels is not valid) supporting the steady-state idea was that it was caused by scattering of light from interstellar dust. However, the physicist George Gamow and others in the middle of the present century showed that the natural abundance of elements in the universe could be accounted for by the existence and expansion of an undifferentiated state of matter. This gave great weight to the idea of an evolving universe (Gamow, 1945, 1952), and this view has now become generally accepted. All but absolute proof for it was brought forward by Arno Penzias and Robert Wilson (1965) (winning the Nobel Prize for it in 1978) at Bell Telephone Laboratories through the discovery of cosmic microwave radiation with a temperature equivalent of $2.7°K$ (so-called "background radiation") that is predicted (Dicke $et\ al.$, 1965) by theoretical physics to have resulted from the processes inherent in the universe's explosive origin ("big bang"). As discussed in detail elsewhere (Penzias, 1978), analysis of these processes also leads to the conclusion that neutrons would be found early (first few moments) during the expansion in equilibrium with about equal amounts of protons ("pair production"). After a further expansion and cooling, pair production would cease, combination of protons and some neutrons would lead to deuterons, while other neutrons would decay to protons and electrons. Most of the deuterons formed would subsequently decay by photo-induced disintegration, but some should remain. Current studies of the cosmic distribution of deuterium offer another way to examine the "big bang," and they are at least consistent with it. In addition, the critical density of matter needed to close the universe (constant volume) is 2×10^{-29} g/cm^3. The deuterium studies lead to an actual value (4×10^{-31}) in good agreement with the value derived from kinematics of galaxies, and this number is substantially below the critical value. The subcritical character of the density thus is consistent with a continuation of expansion and certainly means the universe cannot be in a process of contraction. Viewed in another way, the universe itself has evolved and continues to evolve.

Velocity–distance relationships in the expanding universe have made it possible to calculate that we are presently about 17 billion years away from the "Ylem," as Gamow has called the primordial state of matter. A similar number has been obtained by a very different method. Meteorites (extraterrestrial rocks that strike our earth) have been dated by radioactive procedures. In 1977 Schramm and Hainebach, using the decay of rhenium, found a meteorite to be 20 billion years old. Other workers reported a value of 14 billion years using uranium the year before. Even though none of these methods is completely accurate, the fact that the numbers derived in different ways and by different people are of the same order of magnitude adds still further weight to the assumption of a cosmic beginning occurring in the vicinity of 20 billion years ago.

C. THE AGE OF THE EARTH

Numbers of a similar size to those for the universe, though a bit smaller, have been obtained for the age of our own planet. About 200 years ago it was recognized from the temperature gradient in German mines that the earth is cooling. Subsequently it was discovered in several ways (e.g., seismic) that the earth has an approximately 30-km crust of granite with a high content of SiO_2, under which with an increasing temperature is a very much thicker SiO_2-deficient mantle covering an inner core of molten iron or iron–nickel (Table 3). The undisturbed crustal rocks are relatively easy to find in several places and have been dated as 3.3–3.4 billion years old on, for instance, the Kola Peninsula. The rocks must be undisturbed, incidentally, because one must be able to find both a given radioactive nuclide and its decay products in order to know the entire amount of material which at some point in time, e.g., when the last melting and consequent mixing occurred, was present in an undecayed state. From the ratio of radioactive nuclide to its decay products together with the rate of decay ($t_{\frac{1}{2}}$) the elapsed time is calculable. Less easy to find are rocks which protrude through the crust from the mantle. Since cooling and solidification of the crust would have occurred earlier than in the case of the mantle, the latter should be the older. Indeed, such is the case, as shown, for example, with St. Paul's Rocks in the South Atlantic Ocean, which have been found to be 4.5 billion years old. The earth is believed to have arisen not as a captured object but integrally with the evolution of the sun as it became a discrete hot star. Thus, the earth, and presumably the solar system, must be 5–11 billion years old, representing something like half or more of the age of the universe. Refinements of these age relationships will probably follow, but the general size of the numbers seems to be secure.

Table 3. Structure of the Earth

		Depth (km)	Density (g/cm^3)	Temp. (°C)	Pressure (atms.)	Age after solidifying (10^9 years)
		0	3.5	20	0	
Lithosphere {	Crust	40				ca. 3.3
		100				ca. 4.6
Asthenosphere {	Mantle Mg- and Fe-rich silicates	250				
		1500	4.5	2500	0.8	
		2900	5.5	3500	1.5	
	Core Fe–Ni					
		6370	12.5	4200	3.7	

D. CONTINENTAL POSITION AND AGE

The ages and position of the continents are quite important to our understanding of the changes in life forms that have occurred, because during the time when fossils were being deposited the continents were changing their positions. It is not entirely clear when the major land masses made their first appearance, but rocks in the United States, Europe, and Russia are 200–600 million years old and parts of South Africa and Canada are at least 1 billion years old (Table 4). Some smaller land masses are much younger. The surface of Olduvai Gorge, where the Leakeys found hominid fossils, is only 2 million years old. Darwin's famous Galápagos Islands rose from the sea just a little earlier (3 million years ago) (Adams *et al.,* 1976). Volcanism also can be relatively recent. Thus, the magmatic system in Yellowstone National Park is 2 million years old, probably constituting a plume through the top layer of the earth (Eaton *et al.,* 1975). At the extremely short end of the scale are volcanic islands that have formed and disappeared beneath the v ives in the lifetime of a single human being. It is the ages and positions of the great continents, however, which are of the greatest interest.

Table 4. Some Dated Rocks from Various Strata

Sample	Place	Method	Age (years)	Period
Biotite	Olduvai Gorge	K–Ar	2.0×10^6	Pleistocene
—	Galápagos Islands	K–Ar	3×10^6	Pleistocene
Pitchblende	Colorado	^{238}U–^{206}Pb	5.9×10^7	Paleocene
Biotite	New Jersey	K–Ar	1.95×10^8	Triassic
Biotite	England	Rb–Sr	3.95×10^8	Devonian
Glauconite	USSR	K–Ar	6.10×10^8	Cambrian
Cherts	S. Africa	—	3.1×10^9	Archean
—	St. Paul's Rocks	—	4.5×10^9	Origin of
Galena	Ontario	^{238}U–^{206}Pb	4.8×10^9	planet
—	Meteorites	Various	$0.46–2.0 \times 10^{10}$	Origin of solar system

We can assume from the datings (Table 4) that most of the dry land rose above what is now the ocean floor 100 million to 1 billion years ago. Unfortunately, nothing is known of the positions of these early continents. What has happened in the intervening time is better understood (for keys to the literature, see Dietz and Holden, 1970; Molnar and Tapponnier, 1975; Crawford, 1974; Wegener, 1966; Hallam, 1975). It has to do with the movement of masses of lithosphere over the asthenosphere ("plate tectonics"). Each mass—called a plate—carries a continent on top of it. The movements of these plates ("continental drift") were occurring prior to 300 million years ago and continue to occur today, causing earthquakes and volcanism. Our knowledge of this subject stems from the ingenious ideas of Alfred Wegener in Germany and G. B. Taylor in America during the 1920s. Modifications were made by Harry Hess and Robert Dietz in the 1960s, and the discovery and detailed investigation of the mid-Atlantic ridge and the spreading of the ocean floor as well as other careful studies have placed the existence of plate tectonics on a firm foundation. Although the reasons for the movements are still a matter of speculation (Kerr, 1978), the continental plates are known to be separating on one side and colliding on the other. Magma upwells in the spreading areas, and compression and overriding occurs where the plates collide. Dating of the ocean floor has been a powerful tool with which to authenticate this. In the Atlantic Ocean, for instance, current upwelling of magma through a fissure spreads new rock (after cooling) between North America and Europe. Dating of the ocean floor has proven that the youngest section is the mid-Atlantic Ridge, which contains the fissure. The dates (determined by reversals in magnetism) are progressively older as the distance from the ridge gets greater on either side.

The chronology of plate tectonics has been reconstructed as follows.

Prior to 300 million years ago the events remain hazy, but collision of North America with Europe, producing the Appalachian Mountains, probably occurred in the Devonian or Silurian epochs. These epochs are in the oldest half of the Paleozoic era and are coincident with a fossil record of life that had already evolved to highly organized multicellular forms as complex as fish. By 300 million years ago three great masses of land were thought to exist. One of them was a composite of North America and Europe and straddled the equator. To the northeast was Asia, separated from Europe by what is called the Ural Sea. South of Asia across a stretch of ocean named the Tethys Sea lay a mass of land (Gondwanaland) roughly equal in size to the other two. Gondwanaland comprised what is now South America, Africa, Antarctica, Australia, and India. The Atlantic Ocean separated Gondwanaland from North America-Europe. During the next 75 million years the three early continents coalesced into a gigantic entity, called Pangaea by Wegener, stretching from the Arctic to the Antarctic. The Tethys Sea was closed on the west, forming the start of the Mediterranean Sea; the collision of Europe with Asia eliminated the Ural Sea and produced the Ural Mountains; and the Atlantic Ocean had all but disappeared. The outlines of Pangaea, composed of Laurasia (North America-Europe-Asia) and Gondwanaland, in the early Trassic epoch about 225 million years ago are now reasonably well defined as a result of computer fitting of the true edges (2000 m below the present level of the ocean) of the existing continents (Dietz and Holden, 1970).

Pangaea then proceeded to break up, splitting through the middle by a sort of clockwise rotation of Laurasia, with a pivot at Gibraltar, and by a counterclockwise rotation of Gondwanaland. At the same time, breaks occurred at the bottom of Gondwanaland splitting off Antarctica-Australia and India as separate pieces. The counterclockwise movement drove the latter fragments northeast and, coupled with the clockwise movement of Laurasia, opened up the beginnings of the Caribbean Sea and the North Atlantic Ocean. By the Triassic epoch 180 million years ago, then, Laurasia remained intact, and South America was still joined to Africa, but India and Antarctica-Australia were separate and moving northward. By 150 million years ago (Jurassic) South America had begun to move away from Africa, and North America from Europe. These separations continued to increase, and by 70 million years ago a crack had occurred between Antarctica and Australia and between Africa and Madagascar. The present configuration of the continents was being approached at this time, except that North America was still joined to Europe through what was to become Greenland, the Tethys Sea was still open to the east, India had not yet collided with Asia, and Australia was still substantially southwest of its current position. During the final 70 million years, bringing us to today, South America and North America came close enough for volcanism to form a land bridge (Central America), collision of Africa and India with Asia occurred, forming the Himalayas; North America and Europe became completely separated; and in the northern part of the mid-Atlantic enough magma poured forth to form Iceland.

In summary, overall motion of the continents since formation 230 million years ago of Pangaea on one side of the planet extending from pole to pole has resulted in three major types of cleavage distributing the land mostly through the northern hemisphere and around the globe. One cleavage was almost exactly equatorial through the Laurasia–Gondwanaland suture. The second was approximately an axial one with an equatorial displacement such that fracture occurred in the northern hemisphere 10–20° east of 0° latitude and 10–20° west of it in the southern hemisphere. The two western quadrants (North and South America) moved northwestward, and the two eastern quadrants (mostly Asia and Africa) moved northeastward. There was a gain in latitude of some 20° for most of the land and a longitudinal expansion from 180° (60° west to 120° east) to essentially 360°. In addition, India moved as much as 60° north (from 40° south latitude to 20° north latitude) with Australia moving half that far. Antarctica alone seems to have remained more or less in its same position except for some movement toward the south pole.

As a result of changes in the positions of the continents, the patterns of weather undoubtedly also changed drastically, which in turn must have shaped the environment of life at a given time and place. We know nothing directly about such matters except that the general drift of the continents northward presumably cooled North America and Asia during the last 300 million years while somewhat warming Africa and South America. Climatic change in India and Australia must have been the most dramatic of all with a profound warming as they travelled away from the pole.

E. DATING OF BIOLOGICAL EVENTS

Biological artifacts of previous times have been dated by a variety of methods, and can be grouped as direct and indirect. The latter for the most part consist of dating a specimen found in a geologic layer by using radioactive techniques with inorganic atoms (Tables 1–4) for dating either the layer itself or the layers above and below it. Extensive correlations have now been made by geologists and paleontologists such that in some cases a stratum can actually be dated simply by its physical and biological characteristics without resorting to primary dating. Dating of strata is useful for events occurring more than 100,000 years ago.

The direct method for dating is applied to recent events (<100,000 years ago), and the biological artifact itself is examined. In the 1940s Willard F. Libby pioneered dating with ^{14}C. The method depends on the assumption that ^{14}C is created in the upper atmosphere by bombardment with cosmic rays and that a steady state was long ago reached between the rate of formation and the rate of radioactive decay ($^{14}C_6 \longrightarrow e + ^{14}N_7$). Under these circumstances, when an organism is alive, it will be in equilibrium with the environment containing the constant amount of ^{14}C (more precisely the constant ratio of ^{14}C to ^{12}C). At death, equilibrium is terminated, and the pool of ^{14}C will disappear according to the first-order decay law with a $t_{\frac{1}{2}}$ of 5700 years. An elegant way to check the validity of this method was recently developed by Hans E. Suess through the use of tree rings that permit a direct digital count of elapsed time in terms of years. The assumptions of a steady state in the $^{14}C/^{12}C$ ratio were proven correct except for a small correction. Suess found that beyond 1000 B.C. the ^{14}C dates are increasingly less than the actual dates, the discovery being made first with the bristlecone pine, the oldest of living things. A living specimen can be accurately dated to its origin several thousand years ago. Through overlap of patterns in tree ring sequences among living and dead trees, it has been possible to develop an accurate dating technique back to 6270 B.C., and by comparison with ^{14}C dates to calibrate the latter technique. For instance, uncorrected ^{14}C dates of 1600 B.C. and 4100 B.C. actually represent, respectively, dates of 2100 B.C. and 5000 B.C. Radiocarbon dating is limited by the length of $t_{\frac{1}{2}}$ (5700 years), the sensitivity of detection of ^{14}C, and the size of the sample required (grams). If enough sample is available, its age can be determined with reasonable accuracy over several half-life periods, with a limit of about 50,000 years. The recent advent (Muller, 1977) of the use of the electrostatic accelerator (cyclotron) for the determination of ^{14}C (sort of a mass spectrum) opens the door to the dating of much smaller samples (3–15 mg), which should be of great interest to biochemists, who usually have very little material (Muller *et al.,* 1978; Bennett *et al.,* 1978). Graphite has been dated, for instance, as 48,000 ± 1300 years old on a 15-mg sample. The accelerator technique also extends the dating range to about 100,000 years.

Another technique that allows recent events to be dated is independent of radioactivity. While not directly applicable to lipids, it is nevertheless intriguing and involves the rate of racemization of L-amino acids (Masters and Zimmerman, 1978, and references cited). Aspartic acid with a half-life at 20° of 15,000 years has one of the largest rate constants. Since the rate falls essentially to zero at very low temperatures, even the age at death of an ancient specimen can be determined with tissue, such as bone, with a low turnover rate during life. When the original formation of a tooth, for instance, is taken as time zero, it is found that the L/D ratio of enantiomers of aspartic acid in the tooth will decrease during the life span of a living individual. If the person is frozen at death and subsequently examined, the L/D ratio will then be a measure of life-span more or less regardless of when death occurred. In this way an Eskimo woman who was buried alive, probably in a landslide, 1600 years ago (^{14}C and cultural dating, the latter from tatoos) on St. Laurence Island, Alaska, was preserved in a frozen condition. Morphological examination indicated an age of 50–60 years old. Racemization analysis of a tooth gave the age of death as 53 ± 5 years old (Masters and Zimmerman, 1978)! This is a dramatic example of the utility of a technique that would seem to be amenable to many more dating applications and should see increasing use. Another application of the method has been to show that Paleo-Indians were in California 50,000 years ago (Bada *et al.*, 1974). With amino acids, which have small rate constants, it is thought the technique can be extended to events occurring as many as 1 million years ago.

Among recent events that might be dated and that would be of evolutionary interest is the manner in which the central nervous system has changed. Here lipid biochemistry might well have an opportunity. What precise changes, for instance, have occurred in the lipid constitution of the myelin sheath during the evolution of hominids? Judicious use of lipid analysis and ^{14}C and racemization dating might well tell us a lot, especially when compared to values obtained from other animals. It is perhaps worth pointing out that the ^{14}C methods could also have great utility in biochemical ecology, a subject which almost certainly has evolutionary implications. What, for instance, is the "life" of a molecule of sterol or fatty acid following its biosynthesis as it passes through the food chain? Many types of organisms assimilate preformed lipids. Examples are yeast (operating anaerobically or semianaerobically in a deep pool), certain protozoa such as *Tetrahymena* sp., and apparently all arthropods. An exacting analysis of the ecology of lipids could lead to a better understanding of how and why insects, for instance, came to have a block in sterol formation. Is it their phylogeny—that is, are they phylogenetically determined (by having lost or never possessed the pathway) to have to ingest sterol or cease to exist—or was it just an ecological choice in which sterol was easy to get from other organisms leading to disappearance of the endogenous machinery to make these lipids?

CHAPTER 4

The Paleontological Record

A. EXTRATERRESTRIAL LIPIDS AND OTHER MOLECULES

1. Cosmic Atoms and Molecules

There has been a great deal of speculation in the last few decades about the possible spontaneous origin of life either on this planet or elsewhere in space. We shall therefore begin our examination of the past by reviewing what we know about extraterrestrial lipids and other compounds. Early in the present century spectroscopic examination of light reaching our planet revealed that about 99% of the matter of the universe is in the nuclear form corresponding to hydrogen and helium. In the remaining tiny fraction are found all the other elements, including those from which life is constructed (Table 5). After hydrogen and helium, due to the nature of the "big bang," the next most abundant atoms are the next simplest ones, i.e., those of the first row of the periodic table, including the biochemically important carbon, oxygen, and nitrogen.

The existence of interstellar aggregates of atoms (molecules) was recognized as early as the late 1930s and early 1940s with the discovery of CN, CH, and CH^+ by optical spectroscopy. While this method continues to be useful, the later development of radioastronomy is what allowed a great upsurge in discovery of molecules after 1965, as discussed in detail by Gammon (1978). No less than fifty oligoatomic aggregates have now been identified either as radicals or more often as common molecules (Table 6). The greatest number (three-fourths of them) contain carbon, with carbon monoxide leading the list in molecular abundance after H_2 (Table 7). This molecule (CO) is in fact thought to represent about 10% of all cosmic carbon and is about ten times more abundant than water, which is the next most abundant molecule. In general, abundance declines with complexity, but only

Table 5. Cosmic Abundance of Some Elements
of Biochemical Interest Relative to Hydrogen[a]

Element	Relative abundance
H	1.00
O	7×10^{-4}
C	3×10^{-4}
N	9×10^{-5}
Fe	4×10^{-5}
Si	3×10^{-5}
Mg	3×10^{-5}
S	2×10^{-5}
Ca	2×10^{-6}
P	3×10^{-7}

[a]From Gammon (1978).

Table 6. Some Molecules Found in Space[a]

Formula	Name
CO	Carbon monoxide
CS	Carbon monosulfide
NH_3	Ammonia
H_2O	Water
H_2S	Hydrogen sulfide
SO_2	Sulfur dioxide
CH_4	Methane
CH_2NH	Methanimine
HCN	Hydrogen cyanide
HNC	Hydrogen isocyanide
NH_2CN	Cyanimide
H_2CO	Formaldehyde
H_2CS	Thioformaldehyde
$HCOOH$	Formic acid
CH_3OH	Methanol
$HCCH$	Acetylene
$HCCCN$	Cyanoacetylene
CH_3CN	Methyl cyanide
H_2CCO	Ketene
CH_3CHO	Acetaldehyde
CH_3NH_2	Methyl amine
$HCOOCH_3$	Methyl formate
CH_3OCH_3	Dimethyl ether
CH_3CH_2OH	Ethanol
CH_3CH_2CN	Ethyl cyanide
$H(C_2)_6CN$	Cyanotriacetylene

[a]From Gammon (1978).

Table 7. Cosmic Abundances of Some Molecules
Relative to H_2 [a]

Compound	Relative abundance
Hydrogen as H_2	1.00
Carbon monoxide	10^{-4}–10^{-5}
Water	10^{-5}–10^{-6}
Ammonia	10^{-6}
Hydrogen cyanide and isocyanide	10^{-6}
Formaldehyde	10^{-8}
Methanol	10^{-8}
Methyl formate	10^{-10}
Ethanol	10^{-10}
Dimethyl ether	10^{-10}
Cyanopolyacetylenes	10^{-11}–10^{-12}

[a]From Gammon (1978).

slowly, suggesting that still more complex molecules may yet be discovered. Since interstellar molecules are immersed in a field of radiation (especially ultraviolet starlight), the duration of their existence is believed to be only of the order of magnitude of 100 years, but in dust clouds where 95% of the starlight can be blocked the half-lives are estimated to be perhaps 1–100 million years, depending on the molecule. Carbon monoxide is among the more stable.

2. Molecules on the Moon

On the manned flight of the spacecraft Apollo 11 to the moon, lunar samples were returned from the Sea of Tranquility to several laboratories on earth and examined (Meinschein *et al.*, 1970; Burlingame *et al.*, 1970; Murphy *et al.*, 1970; Ponnamperuma *et al.*, 1970). The carbon concentration was 200 ppm, generally in the form of carbon monoxide (compare the interstellar occurrence in Section A-1). No alkanes, aromatics, isopentenoids, fatty acids, amino acids, sugars, or nucleic acids were found. Traces of organics, such as porphyrins, were detected but are regarded as contaminants from the rocket exhaust.

3. Molecules in Meteorites

During the last 170 years at least 40 meteorite impacts have been observed, and in some cases part of the meteorite survived the crash to a sufficient extent to allow chemical and other studies. The origin of meteors and meteorites is not fully understood. Among the possibilities being considered are comets and collision of asteroids in the asteroid belt between Mars and Jupiter, with the fragments assuming orbits that intersect with our own (McCall, 1973; Wetherill, 1979). Various pieces of condensed matter in the solar system with diameters in excess of 0.1 km and with orbits crossing our own are called "Apollo objects," named after the first one (Apollo) to be recognized, in 1932. Some of these are quite sizable, e.g., "1978 SB," which is 8 km in diameter, and about once a century an Apollo object comes as close to us as the moon. In 1937 Hermes, 1 km in diameter, passed us at only twice the moon's distance. Since Apollo objects are so large that they disintegrate on impact, the smaller bodies (meteorites) are what interest us. Some, such as Allende, are believed to be the least-altered samples of presolar material. The so-called carbonaceous meteorites contain carbon other than in its free form of graphite or diamond. They are categorized according to their carbon content as C1, C2, and C3 (Wiik, 1956). Both the amounts of carbon and of water fall as the class number increases and the density rises. Some of the carbonaceous meteorites are listed in Table 8.

Although complex organic molecules were not found on the moon (Section A-2), quite the reverse has been reported for several meteorites, such as Allende, Murchison, Orgueil, Murray, Cold Bokkeveld, Felix, Mighei, Moika, Lance, Warrenton, Homestead, and Holbrook. The first four have perhaps attracted the greatest attention. The Allende and Murchison meteorites are very recent, both having fallen in 1969, Allende on February 28 in Mexico and Murchison at 11 A.M. on September 28 near Murchison, Victoria, Australia. The parent of the latter object broke up during flight and scattered fragments over 5 square miles. The fragments possessed networks of deep cracks extending into their interiors (Kvenvolden el at., 1970).

Before reviewing the results, we must emphasize that a great deal about this subject is unclear, and considerable controversy exists among the workers in the field. The problem is exacerbated by the very thing that permits it to be examined, viz., sensitive micromethods of analysis. Modern methodology allows such small amounts of compounds to be detected and examined in great detail that the question of contamination is constantly present. In our own laboratory, for instance, it is nearly impossible to have cholesterol-free samples of anything. Cholesterol is "in the air" and has been detected in trace amounts (μg) by gas–liquid chromatography in biological samples we know for other reasons cannot have produced cholesterol (see, for instance, Nes et al., 1971). In another laboratory, where sterols in sea water were being studied, the level of cholesterol in commercial sodium hydroxide was sufficiently high to require the investigators to make their own sodium hydroxide for saponification. The situation with other biochemicals is similar. Micromoles of taurine and urea and a large variety of amino acids have been

Table 8. Some Carbonaceous Meteorites[a]

Name	Descent Place	Descent Date	Weight (kg)	Type
Tonk	India	1911	0.01	C1
Alais	France	1806	0.26	C1
Ivuna	Tanganyika	1938	0.70	C1
Orgueil	France	1864	11	C1
Revelstoke	Canada	1965	0.001	C1
Nawapali	India	1890	0.06	C2
Santa Cruz	Mexico	1939	0.05	C2
Cold Bokkeveld	S. Africa	1838	4	C2
Nogoya	Argentina	1879	2.5	C2
Erakot	India	1940	0.11	C2
Mighei	USSR	1889	8	C2
Haribura	India	1921	0.32	C2
Boriskino	USSR	1930	1.17	C2
Crescent	U.S.	1936	0.08	C2
Bells	U.S.	1961	0.3	C2
Murray	U.S.	1950	12	C2
Al Rais	Saudi Arabia	1957	0.16	C2
Murchison	Australia	1969	82.7	C2
Bali	Cameroons	1907	0.01	C3
Kaba	Hungary	1857	3	C3
Moika	New Zealand	1908	4	C3
Vigarano	Italy	1910	16	C3
Brosnaja	USSR	1861	3.3	C3
Allende	Mexico	1969	1000	C3
Karvonda	Australia	1930	42	C3
Felix	U.S.	1900	—	C3

[a]From Mellor (1972).

identified from a single thumbprint (Hamilton, 1965). When the detailed distribution, for instance, of amino acids found earlier in Orgueil (Kaplan *et al.*, 1963) is examined in light of Hamilton's work, it is probable that the results reflect contamination after the meteorite struck the earth. Similarly, the discovery of polycyclic aromatics in Orgueil (e.g., Bitz and Nagy, 1966) and other meteorites has to be considered in light of the fact that polycyclic aromatics in shallow terrestrial sediments have recently been shown to be almost certainly the result of atmospheric pollution from the burning of coal. The timed sequence of their deposition in the sediments over about 100 years follows the known curve for the use of coal as a fuel, decreasing, for instance, after coal was supplanted by oil. These aromatics must also have settled out on meteorites. The question of chirality in meteoric organics is an especially interesting case. Optical activity has been observed by several workers.

Among them are Meinschein *et al.* (1966), who showed a positive Cotton effect centered at about 340 nm with organics from the Homestead and Holbrook meteorites, which are noncarbonaceous chondrites. The organic matter of consequence could be obtained only from the surface. After showing that there was indeed optical activity, they examined several terrestrial rocks including six ranging from Eocene to Precambrian in geologic age that had been carefully collected and stored for later research. A seventh sample had been used for teaching purposes and had been handled by students for many years. All samples yielded some organics, but only one of the extracts was optically active. It was the one handled by students, and the Cotton effect was practically indistinguishable from the one found in extracts of the meteorites. Contamination can actually occur quite rapidly. Less than one week after the Pueblito de Allende meteorite struck Mexico, a sample was taken for analysis. Painstaking attention to the problem of contamination led the investigators (Han *et al.*, 1969) to believe that only 0.1% of the organics present could have been indigenous to the meteorite. Finally there is the problem (which we have not found discussed) of the presumably fiery state of the meteorites as they entered our atmosphere.

Despite the many questions about validity, some of those concerned with the subject believe that at least a portion of the meteoric organics were really present prior to entry into our atmosphere. Thus, the indigenous nature of the alkanes in at least Murray, Murchison, and Orgueil is believed supported by four lines of evidence, as reviewed recently by Anders *et al.* (1973): (1) absence of the isopentenoidal pristane and phytane found frequently in terrestrial samples; (2) presence of light hydrocarbons; (3) low amount of alkanes in Allende (C3), which is metamorphosed and can be regarded as a blank; and (4) a larger difference in the $^{12}C/^{13}C$ ratio (60–80 per million) between carbonate and other kinds of molecules of carbon than do terrestrial samples (25–30 per million). Still, we advise caution in drawing conclusions from the survey that follows.

Following the famous Miller–Urey experiments in the 1950s (Miller, 1953; Miller and Urey, 1959) and the speculations inherent in the Oparin–Haldane hypothesis of chemical evolution (cf. Oparin, 1957), a variety of people argued for a spontaneous origin of life, or at least of biochemicals on earth and possibly elsewhere. It was therefore reasonable to look for artifacts of the process in meteorites, and this followed in the early 1960s (Urey and Lewis, 1966; Urey, 1966; Studier *et al.*, 1965, 1966; Hayatsu, 1964, 1965; Briggs, 1963; Bitz and Nagy, 1966; Clayton, 1963; Kaplan *et al.*, 1963). At the same time, the possibility that the carbon (cyanide) in comets might play a role in the formation of biochemicals was proposed by Oro (1961), who also showed that hydrogen cyanide could be converted to adenine at 90° during 24 hr in a solution of ammonium hydroxide (Oro, 1960). There are actually a great many comets (about 40 million of them), and an estimated 100 may have collided with the earth since its formation. However, no material clearly known to be cometary has been available for collection and study. More recently, with frequent attention to the problem of contamination, several groups of workers have continued the investigation of meteorites, and they report the presence of hydrocarbons of several kinds, isopentenoids, fatty acids, amino acids, and N-heterocycles of the sort found in DNA and RNA. We will survey these in the order given. Earlier and more extensive discussions exist (Anders *et al.*, 1973; Oro, 1973; Hayes, 1967), which, along with the original literature, the interested reader should consult for more detail.

Aromatic hydrocarbons of a great variety have been obtained. From Murchison, for instance, Pering and Ponnamperuma (1971) report some two dozen, including naphthalene (with and without methyl substituents), biphenyl, acenaphthene, fluorene, phenanthrene and methylphenanthrenes, anthracene, fluoranthene, and pyrene. Hayes and Biemann (1968) also found such molecules in Murray. Naphthalene was the most abundant, and among other compounds present were anthracene, phenanthrene, alkylindenes, phenylacetylenes, alkylstyrenes, alkylindanes, pyrene, and acenaphthalene. Orgueil, Allende, and Cold Bokkeveld have yielded benzene and alkylbenzenes (Studier *et al.*, 1965; Simmonds *et al.*, 1969).

Alkanes have been isolated from Orgueil, Murray, Murchison, Moika, and others. The amount of the aliphatics appears to increase as the meteoric class number increases from 7 ppm in C1, through 50 in C2, to 100 in C3, while the noncarbonaceous meteorites have a concentration of only 2. Normal alkanes of both odd and even numbers of C atoms dominate strongly over branched and cyclic ones, and the distributions of chain length are different in different meteorites. The peaks in the distribution are frequently at an odd carbon number, e.g., C_{15} for Moika and C_{21} for Murray, but the true distributions are not easy to assess with precision. A surface rinse with benzene–methanol–hexane, for instance, of Murchison showed about equal amounts of n-C_{16} and n-C_{17} and decreasing amounts of n-C_{15}, n-C_{14}, and n-C_{13}, but a benzene–methanol extract of pulverized material showed slight dominance of n-C_{15} in an homologous series decreasing more or less regularly on either side (Anders *et al.*, 1973). The range of chain lengths in the various meteoric alkanes is from about C_9 to C_{29}. Between C_2 and C_8 no n-alkanes have been found except on Allende. A dominance of odd over even chains occurs in some meteorites. Branched chains occur, but in limited variety including 2- and 3-monomethyl and 2,3- 3,4-, and 4,5-dimethyl derivatives. (For details see Nooner and Oro, 1967; Studier *et al.*, 1968, 1972; Simmonds *et al.*, 1969; and Kvenvolden *et al.*, 1970.)

The distribution of alkanes in meteorites is suspiciously similar to that found in terrestrial organisms, especially microorganisms. People who work on meteorites tend to neglect alkanes in extant, living systems. Not only might algae, etc., infect the meteorite after impact, but airborne hydrocarbons from dead cells might act as contaminants. In the latter case, there could certainly be a fractionation as the compounds left the dying cell, entered the air, and then were adsorbed by rock. An extraterrestrial origin of these compounds seems to us not yet a compelling conclusion. Furthermore, as Anders *et al.* (1973) point out, there are empirically some ten thousand structural isomers for a C_{16} hydrocarbon, yet only a half dozen of these have been found. Clearly, whatever process formed alkanes must have had extraordinary selectivity. This very selectivity, however, is what leads Anders *et al.* (1973) to argue for an extraterrestrial synthesis (in the "solar nebula" at 360–400° K and 10^{-6} atm) of the Fischer-Tropsch type (carbon monoxide plus hydrogen and a catalyst), used on earth to produce hydrocarbons industrially. Among the main arguments for the Fischer-Tropsch type of synthesis is that in model experiments a hydrocarbon pattern similar to that found in the C_{15}–C_{16} range on the Murray meteorite is obtained, although such a comparison with the hydrocarbons (C_{11}–C_{20}) of Murchison is much less convincing.

The isopentenoidal hydrocarbons, pristane and phytane found in some algae, bacteria, marine zooplankton, and fish and whale oils (Blumer, 1965; Oro *et al.*, 1967; Han *et al.*, 1969) were reported early in the work on meteorites, apparently confirming the speculations of extraterrestrial life (Nooner and Oro, 1967; Calvin, 1969; Gelpi *et al.*, 1970b; Gelpi and Oro, 1970; Oro *et al.*, 1968; McCarthy and Calvin, 1967). However, Studier *et al.* (1968 and 1972) were unable to find these compounds in extracts of Orgueil or Murray and regard them as contaminants, since they were found only on the surface and not in the interior of Murchison. From Murray and Orgueil, however, camphene and camphor, respectively, have been isolated (Studier *et al.*, 1968).

Fatty acids have been reported by several groups of investigators (Nagy and Bitz, 1963; Hayatsu, 1964; Smith and Kaplan, 1970) in no less than seven meteorites. The dominant chain lengths are C_{16}–C_{18} and include 30% of unsaturates. Although Anders *et al.* (1973) point out that Fischer–Tropsch synthesis can lead to fatty acids, tacitly suggesting an extraterrestrial origin, certainly the distribution of chain lengths and the presence of unsaturates in the samples from meteorites strongly indicates an origin on earth, as Smith and Kaplan (1970) suggest.

Amino acids have been reported present in meteorites both early and more recently. In what is supposed to be an especially secure study, Kvenvolden *et al.* (1970, 1971) isolated 17 racemic amino acids from Murchison, among them glycine, alanine, glutamic acid, valine, proline, and 10 unusual ones, e.g., 2-methylalanine, isovaline, and 2-, 3-, and 4-aminobutyric acid. Other recent studies (Oro *et al.*, 1971; Lawless *et al.*, 1971;

Cronin and Moore, 1971; Hayatsu *et al.*, 1971; Levy *et al.*, 1973; Pereira *et al.*, 1975) have also shown amino acids to be present and have extended the list. The racemic character of the amino acids is generally taken as strong evidence of extraterrestrial formation. Hayatsu *et al.* (1971) believe this conclusion is strengthened by the presence of unusual amino acids and the fact that not only does Fischer–Tropsch synthesis in the presence of ammonia give an array of usual and unusual amino acids but the distribution is similar to the assemblage found in meteorites (also see Anders *et al.*, 1973). The extraterrestrial origin of these compounds is given still further weight by the fact that isovaline lacking an enolizable hydrogen is among the racemic amino acids of Murchison (Pollock *et al.*, 1975). The amino acids are thus thought to have arisen *de novo* in the form of both enantiomers, unlike biosynthesis on earth, where asymmetric induction occurs.

Among the N-heterocycles identified in Orgueil and Murchison are biological types (adenine and guanine) as well as unusual ones, e.g., melamine, ammeline, and guanyl urea (Hayatsu *et al.*, 1975; Hayatsu *et al.*, 1972). Other meteorites are reported to contain heterocycles such as 4-hydroxypyrimidines but none of the usual biological variety (Folsome *et al.*, 1973). Anders *et al.* (1973) regard many of these compounds as derived extraterrestrially by Fischer–Tropsch synthesis.

N-Heterocycles resembling porphyrins have been reported by Hodgson and Baker (1969) in Orgueil, Murray, and others.

B. TERRESTRIAL FOSSILS

1. Molecular Fossils

Although a great deal is known about organismic fossils, recent work has also laid a basis for the investigation of molecular artifacts as well. We shall use the term fossil, whether in an organismic or molecular sense, to imply any remnant of a previously existing form of life from an imprint (such as a footprint or a cast of a tree trunk), through the organized aggregates themselves (shells, fragments of skulls, preserved wood, etc.), to individual molecules either in their original or altered forms. Molecular fossils have been the subject of two excellent earlier reviews (Eglinton and Murphy, 1969; Albrecht and Ourisson, 1971). Two kinds of molecular fossils are found: (1) the unchanged biochemical, and (2) some derivative of the original molecule. In the latter case, for reasons that remain obscure (perhaps bacterial action), reduced forms are frequent. The processes transforming the original to a derivative, e.g., a saturated form, after deposition are known collectively as "diagenesis." Organic molecules are most abundant in material deposited by sedimentation from an aqueous environment. When the sediments are old enough to have formed rock, only about 10% of the organic material can be extracted with solvents. The other 90% is in the form of a polymeric mixture called kerogen. Extractable organics can amount to from 0.1% in rocks to 25% in oil shales. Most of our present knowledge of the rocks is derived from the soluble substances; the information available on kerogen suffers from the fact that kerogen is "solublized" by treatment with a mixture of hydrofluoric and hydrochloric acids, which could cause rearrangements, dehydrations, etc.

n-Alkanes are usually encountered in all sediments, including those from the Precambrian (Oro *et al.*, 1965; Meinschein, 1965), but an interpretation of past events, both biological and geological, from the composition of aliphatics is clouded by complexities not yet sorted out. The composition depends on, among other things, depth, age, and temperature. In some cases, however, very little diagenesis has occurred.

Knoche and Ourisson (1967) examined a clay layer from a sandstone 200 million years old containing organismic fossils only of horsetails (*Equisetum*). The n-alkane distribution (mostly odd, C_{23}–C_{29}, with a maximum of C_{27}) was essentially the same as obtained from an extant specimen of *E. brongnarti* and *E. palustre* (Stransky *et al.*, 1967). Similarly, the Messel oil shale, a 50-million-year-old stratum never covered deeply (< 300 m) and hence not subjected to the higher temperatures and pressures, abounds in fossil flora closely related to those extant today in tropical regions. The n-alkanes in the shale were primarily of odd numbers of carbon atoms, with a peak in the distribution at C_{27} (Albrecht and Ourisson, 1971). When the shallow oil shale of Creveney (180 million years old) of marine origin was investigated, it also was found to bear testimony to the life forms originally present, but in this case in the algae and other microorganisms of the ocean. The distribution of n-alkanes had a strong maximum at C_{17} and ranged from C_{13} to C_{35} with an odd-to-even ratio of about unity, excluding the disproportionate amount of C_{17} (Albrecht and Ourisson, 1971). Johns *et al.* (1966) examined the Green River shale, a 60-million-year-old limnetic (freshwater) deposit. A typical distribution (odd C numbers, peaks at C_{29} and C_{31}) for higher plants was found, except that a strong spike occurred at C_{17}. The former probably represents plants of the shoreline and the latter the lower flora of the sea. On the other hand, correspondence between morphological fossils and molecular fossils is not always clear. Knoche *et al.* (1968) studied a Lower Triassic sandstone of the Vosges in France that had fossils of a conifer now extinct, *Voltzia brongnarti*. The n-alkane fraction was nearly all constituted by the even-numbered octacosane (C_{28}), which is not representative of tracheophytes. Similarly, other sedimentary cases have a slight preponderance of n-docosane (C_{22}). These even-numbered alkanes may have their origins diagenetically in the corresponding fatty alcohols. The n-alkanols of plants usually are of an even number of carbon atoms.

In an investigation of the relationship of depth to the distribution of chain length, Philippi (1965) found that in Miocene deposits in California above 2400 m there is a dominance of odd-numbered chains (C_{27}–C_{33}), but below this depth the preponderance decreases until it is no longer evident beyond 3300 m, presumably as a result of diagenetic formation of even-numbered chains in the older and deeper sediments. Analogously, chert from the 2-billion-year-old Gunflint iron formation off the north shore of Lake Superior in Canada has yielded n-alkanes with a range of C_{16}–C_{32} and two maxima (C_{18}–C_{19} and C_{22}) and no predominance of odd or even chains (Oro *et al.*, (1965), and the n-alkanes of the Fig Tree Series in South Africa, dated at over 3.2 billion years old, show a 1 : 1 ratio of odd and even chain lengths (peak at about C_{22}–C_{24}) (Calvin, 1969).

Fatty alcohols and acids have been found in a number of sedimentary formations. The Green River shale, for instance, contains n-docosanol (C_{22}) as the principal alcohol (Sever and Parker, 1969). Fatty acids of even chain lengths (especially C_{16} and C_{18}) dominate in younger sediments, with an even-to-odd ratio >4, while odd-numbered ones become important in the older sediments, with an even-to-odd ratio near unity (Kvenvolden, 1967; Cooper and Blumer, 1968; Parker, 1969). Unsaturated fatty acids are also found (Parker, 1969). The amount is greater in recent sediments, but $C_{16:1}$ and $C_{18:1}$ have been identified in the Eocene Green River shale and the Cretaceous Thermopolis shale. Even in the Alum (500 million years old) and McMinn (1.6 billion years old) shales unsaturates can be detected. Das and Smith (1968) showed, however, that the percentage of $C_{16:0}$ and $C_{18:0}$ acids increases at the expense of the corresponding unsaturates as the age of the sample increases from zero through tens of thousands of years to hundreds of millions of years old (Table 9). The study by Das and Smith seems to be a good illustration of diagenetic reduction, but their data (Table 9) also show, for reasons not evident, a rise in the amount of $C_{14:0}$ as one proceeds from extant algae to the Triassic fossils (200 million years old), with the still older fossils showing a drop again toward the value for the current algae. The amounts of unsaturated C_{14} acids cannot account for this. Is the $C_{14:0}$ acid derived from a hydrocarbon? Did the ancient algae have more saturated acids than extant ones? More research is clearly needed.

Isopentenoids occur in one form or another in sediments from very recent times on back to early Precambrian deposits. As with the n-alkanes, the extent to which the stratum has been altered by pressure and temperature affect whether or not the original biochemical will be found unaltered. β-Carotene has been identified in the Searles Lake sediment, believed to be 20,000 years old, but the oldest of perhaps six unsaturated carotenoids identified is only

Table 9. Fatty Acids in Fossil Algae[a]

Sample	Age (years)	\% in mixture with chain length indicated[b]										
		10:0	12:0	14:0	14:u	15:0	16:0	16:u	18:0	18:u	20:0	Other
Algae	Extant	6	1	6	1	1	33	19	3	28	0	0
Nevada	1×10^4	4	2	10	0.2	3	48	8	15	12	tr	tr
Green River	6×10^7	2	1	17	0.1	4	52	4	16	3	1	tr
Australia	2×10^8	tr	5	18	tr	3	52	2	17	2	0.6	0.3
Germany	3×10^8	tr	6	17	tr	3	53	1	18	2	tr	tr
Arizona	3×10^8	tr	tr	13	tr	2	60	tr	23	1	tr	2
Mexico	6×10^8	tr	tr	11	0	1	68	tr	20	0	0	tr

[a]From Das and Smith (1968).
[b]u = Unsaturated; tr = trace.

100,000 years old at most (see Schwendinger, 1969, for a key to the literature). The fully saturated skeleton (carotane) has been found in the 60-million-year-old Green River shale (Murphy *et al.*, 1967) and is found in petroleum. Changes of both a diagenetic and apparently an ecological nature alter the fate of isopentenoids. Fucoxanthin is the major xanthophyll of diatoms, which are believed to be the major (85%) plant biomass of the oceans. Yet, this carotenoid is not common even in recent sediments. Harvey (1957) has pointed out that algae probably only rarely reach the ocean floor, being eaten instead by zooplankton. An oxidative degradation product (2-methyl-5-dimethyl-cyclohexaneacetic acid) either of biochemical or diagenetic processes of the six-membered ring of carotenoids is found in crude oil (Cason and Graham, 1965). Similarly, sterols probably do not usually last very long. Turfitt (1943) showed that cholesterol is rapidly destroyed biologically in aerated soils. Nevertheless, unchanged individual sterols and triterpenoidal alcohols have been identified in sediments as much as 70 million years old (vide infra), and camphor and borneol are reported to be present in the 2-billion-year-old Ketilidian formation (Calvin, 1969).

Using a colorimetric method Schwendinger and Erdman (1964) showed that sterols are quite common in recent sediments both in limnic and marine environments at concentrations of 60–300 ppm. Later Attaway and Parker (1970) examined a sample derived from the deepest meter of a 3-m "piston core" taken from the bottom of Baffin Bay, Texas. The sample was about 2500 years old. Using gas–liquid chromatography and mass spectroscopy they identified stigmasterol and the homologous series of cholesterols with a 24-H, 24-CH_3, and 24-C_2H_5 characteristic of "main line" tracheophytes. The tracings given of the gas–liquid chromatogram showed additional materials. Although not discussed by Attaway and Parker, the published mass-spectral data (M^+ 386, 388, 400, 412, 414, 416, and 428) indicate the presence of cholestanol (388), 24-ethylcholestanol (416), and a C_{30} compound (428). Dihydrolanosterol and cycloartanol are not likely candidates, owing to their very small concentrations in living systems, and the C_{30} compound was more likely a pentacyclic triterpenoid, e.g., dihydro-β-amyrin. Attaway and Parker (1970) also identified stigmasterol and the three homologous cholesterols from a "grab" sample taken from the San Pedro Basin off the coast of southern California. The tracing given of the gas–liquid chromatogram again showed other components. No mass-spectral data were given. The lipid yield from the Baffin Bay sample was 400 ppm on the basis of dry weight, and the sterols were 6 ppm. The San Pedro sample had a much higher content both of lipids in general (10,000 ppm) and of sterols (40 ppm). Sitosterol has also been reported to be present along with sitostanol in peat moss (Ives and O'Neill, 1958), "young" peat (Ikan and Kashman, 1963), as well as in Scottish lignite from the Oligocene period (Ikan and McLean, 1960). The Green River shale is reported to contain C_{27}, C_{28}, and C_{29} steranes and other hydrocarbons, (Calvin, 1969).

Pentacyclic triterpenoids, perhaps due to their high melting points, actually appear to be the first of the polycyclic isopentenoids to be demonstrated as unaltered molecular fossils. Nearly 50 years ago Ruheman and Raud (1932) crystallized betulin from Central German brown coal. Brown coal is, incidentally, a young, less modified deposit than "regular" coal. The various coal deposits in increasing age and diagenesis are peat, lignite, brown coal, bituminous coal, and anthracite. In the years after the work of Ruheman and Raud many triterpenoidal alcohols have been isolated from coal and some also from petroleum (see Streibl and Herout, 1969, for a detailed discussion and key to the literature), and from oil shale (Albrecht and Ourisson, 1969). In the latter case, the Eocene Messel oil shale has yielded isoarborinol (Figure 1), which occurs today among the Rutaceae, Rubiacea, and Gramineae. Isoarborinol with an HO group at C-3 belongs to the hopane family of triterpenoids. 3-Desoxyhopanes have also been found in a variety of petroleums, shales, and other deposits (Van Dorsselaer *et al.*, 1974). Among the coals rich in triterpenoids is the North Bohemian brown coal ("montan wax"), from the Josef–Jan Mine, believed to be 30–50 million years old. As many as 26 materials melting above 200° have been obtained by Sorm and his colleagues in Czechoslovakia (Jarolim *et al.*, 1965, and references cited). Optical rotations performed on most of them proved chirality had been preserved. For a partial list, see Table 10 and Figure 1. In addition to the alcohols and other oxygenated compounds, hydrocarbons were found. Many of these are clearly derived from triterpenoids. An example is the fully aromatic 1,2,9-trimethylpicene, which has the skeleton of α-amyrin (or ursolic acid) lacking one of the two methyl groups at C-4 in ring A (steroid numbering) and all of the angular methyl groups. Aromatic compounds of this sort can be obtained experimentally by catalytic dehydrogenation (see, for instance, Nes and Mosettig, 1954), and their appearance in sediments almost certainly represents diagenetic alterations of the original biochemical polycycles. The picenes presumably are derived from pentacycles and the phenanthrenes from the tri- and tetracycles (tri- and tetracyclic terpenoids, steroids, and euphoids). As suggested by Streibl and Herout (1969), steroids may also be the source of anthracenes through rearrangement of the original steroid to an anthrasteroid followed by dehydrogenation (Nes and Mosettig, 1954; Nes and Ford, 1963). Skrigan (1951, 1964) has presented direct evidence for such aromatizations in peat. The conversion of abietic acid (a diterpenoid of conifers) to the methyl isopropyl phenanthrene known as retene was demonstrated in wood stumps sunk in peat. Skrigan also showed that simple benzene derivatives can arise in this manner. Younger layers (perhaps 2000

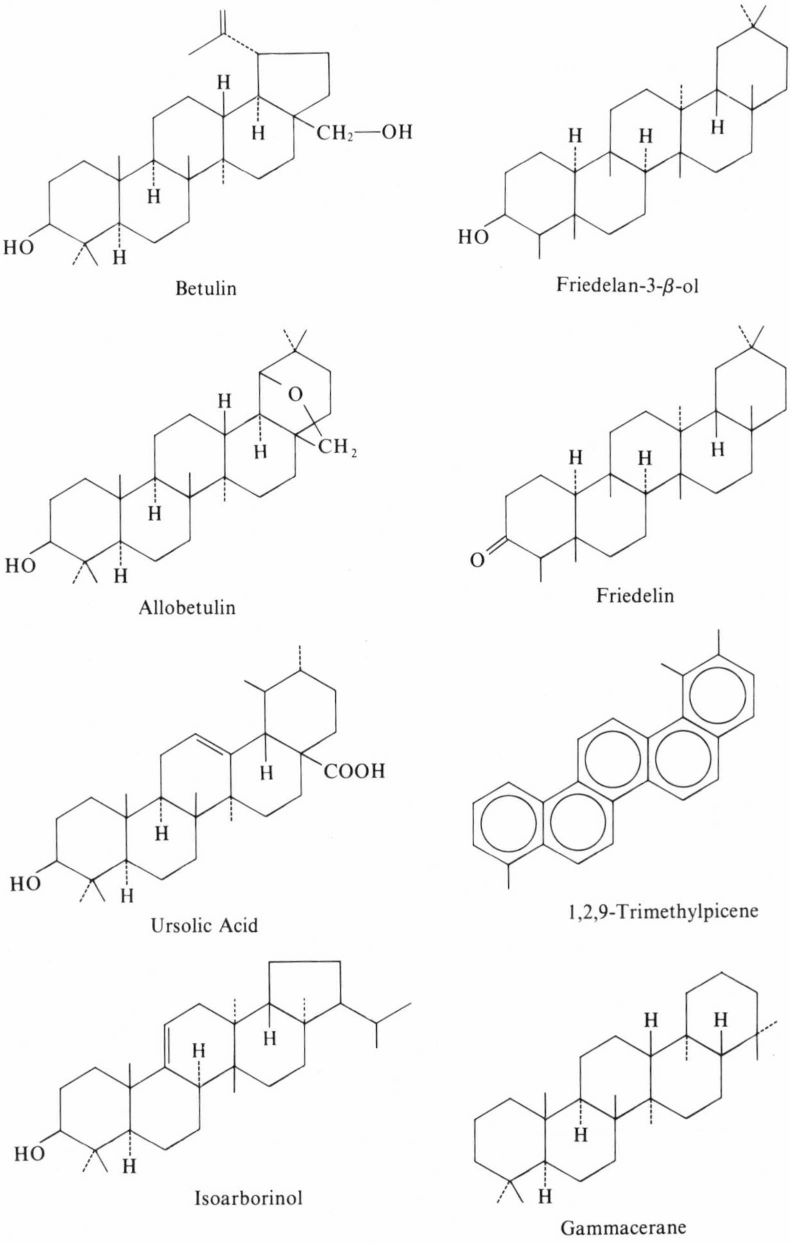

Figure 1. Some pentacyclic molecular fossils which are 40–60 million years old.

Table 10. Some Triterpenoids in Various Sediments

α-Apoallobetulin[a]	Ursolic acid[a]
Allobetulin-2-ene[a]	Tetrahydro-2,2,9-trimethylpicene[a]
Friedelin[a,c,d]	1,2,9-Trimethylpicene[a]
Friedelan-3α-ol[a]	Gammacerane[e]
Friedelan-3β-ol[a,c,d]	Isoarborinol[b]
Oxyallobetul-2-ene[a]	Isoarborinone[b]
Allobetulone[a]	3-Desoxyderivatives in hopane family[f]
Allobetulin[a]	

[a]Coal from Josef–Jan Mine. [d]Scottish peat.
[b]Messel oil shale. [e]Green River oil shale.
[c]Lignite. [f]Petroleum, coals, shales, etc.

years old) of peat contain α-pinene, while in the older (and deeper) layers (7000 years old) the pinene has been replaced by the benzenoid p-cymene and its saturated analog p-menthane (Figure 2). The experimental dehydrogenation of diterpenoids to retene is well established, e.g., the conversion of isosteviol to pimanthrene on Pd–charcoal (Mosettig and Nes, 1955).

Fossil polycyclic diterpenes and their diagenetic products have been the subject of commercial interest giving rise to a good bit of study. Copal, amber, and brown coal are among the forms in which the deposits occur. Among the compounds isolable are fichtelite, retene, and iosene. The latter is a tetracycle known chemically as α-dihydrophyllocladene (Figure 3), and it occurs as such in some extant tracheophytes such as conifers.

Acyclic diterpenoids have also been found as molecular fossils. Of great interest are the apparent fossils of phytol due to its place as a part of the cholorphyll molecule. Cholorphyll itself is never found in old deposits, presumably due to diagenetic changes. However, impressions of green leaves have been found in middle Eocene brown coal (Weigelt and Noack, 1931; see also Blumer, 1950, for porphyrins in bituminous coal). Spectra of extracts of these coals revealed derivatives of chlorophyll. A subsequent study of the brown-coal impressions showed the specific presence of methylpheophorbide-*a* (Dilcher *et al.*, 1970). This derivative of the porphyrin portion of chlorophyll, found in the Neumark–Sud open-pit brown-coal mine in the Geisel Valley near Halle, East Germany, is the oldest (47 million years old) so far discovered. The isopentenoid portion of chlorophyll has been encountered, at least in diagenetically changed forms, in both recent and older sediments, as in the case of dihydrophytol (Sever and Parker, 1969) and phytanic and norphytanic (pristanic) acids (Figure 4), the latter being accompanied by farnesanic and norfarnesanic acids in California petroleum

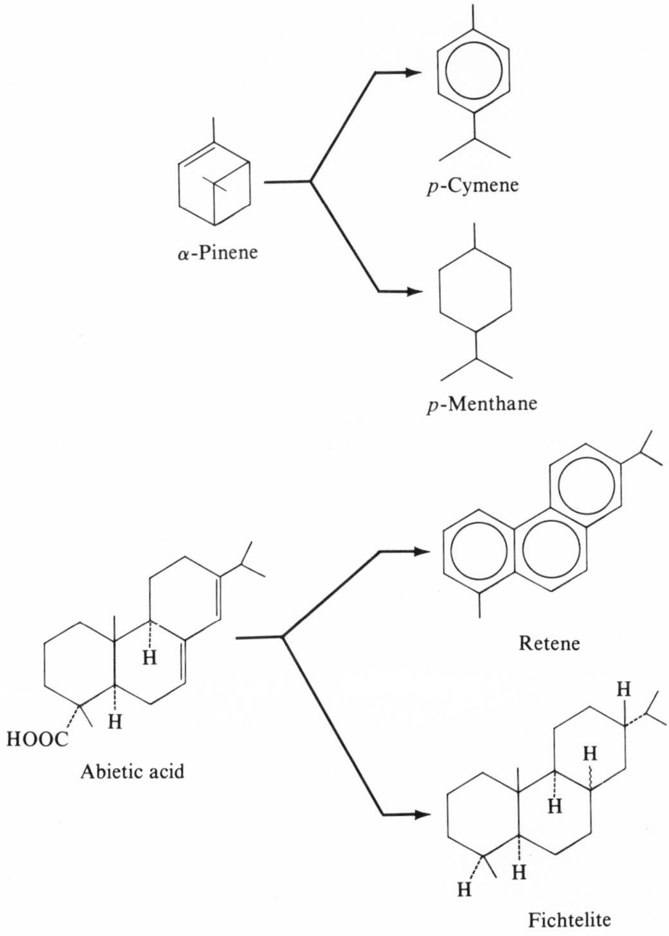

Figure 2. Diagenetic saturations, decarboxylations, and aromatizations occurring in peat during fossilization (Skrigan, 1951, 1964).

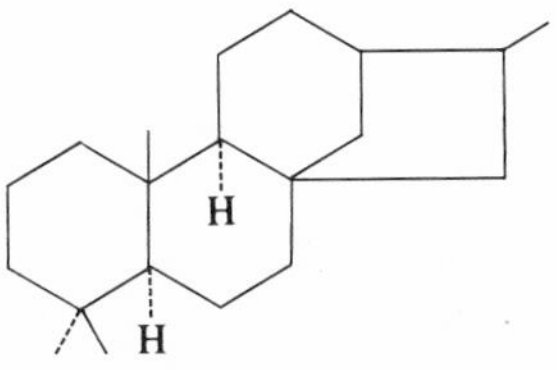

α-Dihydrophyllocladene
(iosene)

Figure 3. Structure of iosene. Found in brown coal and in extant plants, iosene is a reasonable diagenetic product of other diterpenes.

(Cason and Graham, 1965). Phytanic and norphytanic acids have also been shown to be present in the Eocene Green River shale from Sulfur Creek (Eglinton *et al.*, 1966). In fact, more of each of these acids was present in the shale than of any of the array of fatty acids present. Subsequently, Burlingame and Simoneit (1968) found phytanic and norphytanic acids in the "kerogen matrix" of the Green River formation. The Green River shale has also yielded phytane as well as much smaller amounts of pristane (C_{19}) and two other isopentenoidal hydrocarbons each of C_{18} and C_{16} (Calvin, 1969). The Soudan shale in Minnesota (2.7 billion years old) contains pristane, phytane, and a C_{21} isopentenoidal hydrocarbon in decreasing amounts in the order given (Calvin, 1969). The C_{21} compound, present also in the Michigan Nonesuch shale (1 billion years old), has been identified as 2,6,10,14-tetramethylheptadecane (McCarthy *et al.*, 1967). This can be regarded as representing homophytane (1-methylphytane, numbering from the C atom originally bearing the oxygen atom in phytol), i.e., the addition of a C atom to phytane, but more likely it is a fossil of the higher polyprenols (dolichols, solanesol, etc.) and is derived by

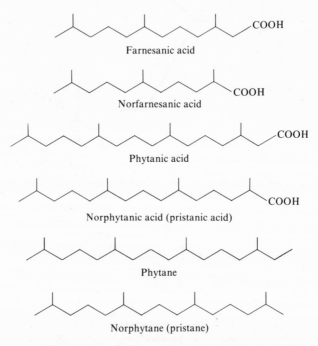

Farnesanic acid

Norfarnesanic acid

Phytanic acid

Norphytanic acid (pristanic acid)

Phytane

Norphytane (pristane)

Figure 4. Acyclic diterpenoids found as molecular fossils.

degradation. The possibility that isopentenoidal hydrocarbons are derived from the phytanyl and biphytanyl moieties of the cell envelopes of bacteria must also be considered.

The origins of the isopentenoidal alcohols, acids, and hydrocarbons is a complex matter. Phytol is the isopentenoid in most chlorophylls of higher and lower photosynthetic organisms, from tracheophytes through algae to prokaryotic bacteria, but in some of the bacteria farnesol replaces phytol. Thus, the C_{15} compound, farnesanic acid, may represent this latter organism, while the C_{20} compounds, phytanic acid and phytane, may represent the other ones, although it is often assumed the C_{20} compounds have yielded the C_{15} compounds. The removal of just one carbon atom, however, to give the nor-compounds (Figure 4) still remains to be explained in a really satisfactory manner, despite the obvious explanation that decarboxylation may have occurred. The presence of pristane in marine animals (zooplankton, notably North Atlantic copepods of the genus *Calanus*, in which it amounts to 1-3% of the body fat; and mammalian oils) as well as in some algae and bacteria is believed to explain the fact that pristane is found in the absence of phytane in recent marine sediments (Blumer and Snyder, 1965) and that it dominates all branched-chain alkanes in several sediments, e.g., the marine Créveney shale (180 million years old), which have not been subject to much diagenetic change (Albrecht and Ourisson, 1971). That the character of the sea making the deposit is probably important is suggested by the dominance of phytane over pristane in the Green River and Messel shales derived from fresh waters.

In summary, fossil lipids have been found in sediments ranging from the present back to the earliest Precambrian times 3 billion years ago (Table 11). While the record is erratic and needs much more clarification, especially in view of varying diagenesis, the following conclusions can be drawn tentatively: (1) ecological as well as diagenetic processes influence the fate of biochemicals after biosynthesis; (2) the older a sediment, the more likely it is that diagenetic hydrogenation or aromatization has occurred, as exemplified by the fossils of fatty acids, terpenes, and phytol; (3) the fossils may reflect whether the paleoorganisms lived in a marine or a limnetic environment, as indicated perhaps by the relative occurrence of pristane and phytane and of *n*-alkanes of various chain lengths; (4) the organisms on earth at an earlier time were composed of the same classes of lipids as those today (*n*-alkanes, fatty acids and alcohols, and acyclic and cyclic isopentenoids); and (5) no obvious biochemical evolution can be documented. The last point obviously emphasizes the need for further research. Required will be more demanding analyses of stereochemistry and a search for more of the complicated lipids as well as more detailed knowledge of diagenesis after deposition and of

Table 11. Some Formations Where Molecular or Other Fossils Have Been Found

Formation	Site	Age (years)	Fossils
Sediments	Various	Recent	Many fossils; little fucoxanthin; pristane but not phytane
Pacific Ocean sediment	California	Recent	Sterols
Gulf of Mexico sediment	Mississippi	Recent	Carotenoids
Gulf of Mexico sediment	Texas	2.5×10^3	Sterols
Peat	Europe and America	7×10^3	Mono- and diterpenoids
Searles Lake sediment	—	2×10^4	β-Carotene
An Interglacial Gytja	—	1×10^5	Carotenoids
2400-meter sediments	California	—	n-Alkanes mostly of odd lengths
3300-meter sediments	California	2×10^7	n-Alkanes equally of even and odd lengths
Josef–Jan Mine	Czechoslovakia	4×10^7	Triterpenoids
Lignite	Scotland	4×10^7	Sterols
Geisel Valley Mine	Germany	5×10^7	Chlorophyll derivative
Green River shale	Colorado and other western states of U.S.	6×10^7	n-Alkanes, n-docosanol, fatty acids, steranes, triterpane, phytanic acid, pristanic acids, phytane, pristane and C_{18}- and C_{16}-1-hydrocarbons
Messel Oil shale	Germany	6×10^7	Tropical flora, n-alkanes, fatty acids, triterpenoids

Creveney Oil Shale	France	1.8×10^8	n-Alkanes (C_{17} plus even-odd series); phytane, pristane, and C_{15}-, C_{16}-, and C_{18}-1-hydrocarbons
Antrim shale	Michigan	2.7×10^8	Homophytane
Petroleum	Various	—	n-Alkanes, isopentenoidal acids and hydrocarbons, triterpenoids, and other
Sonora	Mexico	6×10^8	Fatty acid
Bitter Springs	Australia	0.9×10^9	Prokaryotic and eukaryotic cells
Nonesuch shale	Michigan	1×10^9	Homophytane
Beck Springs	California	1.3×10^9	Eukaryotic cells
McMinn shale	Australia	1.6×10^9	Fatty acids
Gunflint chert	Canada	2×10^9	n-Alkanes and microflora
Witwatersrand	S. Africa	$2\text{--}3 \times 10^9$	—
Soudan iron	Minnesota	$>2.7 \times 10^9$	Pristane, phytane, homophytane, and perhaps microflora
Onverwacht and Fig Tree series	S. Africa	$>3.2 \times 10^9$	n-Alkanes and perhaps microflora

ecological processes prior to deposition. The really crucial evolutionary question remains unanswered, viz., whether and to what extent and when there was an evolution through time in the metabolic processes governing biochemistry. The striking fact that isopentenoids, for instance, were present in the very early phases of the Precambrian period 3 billion years ago indicates that the fundamental biochemical processes in extant organisms were present in the very early organisms, e.g., the bacteria and blue-green algae identified morphologically in this same period (Section B-2). Although carbohydrates are not the subject of this review, it is worth mentioning that they too are found from recent to ancient sediments. D-Glucose and D-galactose in particular have been found not only in recent and in Devonian but in early Precambrian (Coutchiching) formations. That there can be little doubt of the extreme biological sophistication of the organisms on the primitive earth is not surprising, since they were indeed living cells.

2. Organismic Fossils

It would be out of place to review the record of organismic fossils in depth here, but there are some essential points worth summarizing, especially in view of recent discoveries. In the first place, unlike what the record of molecular fossils can show us so far (Section B-1), organismic fossils unequivocally document that a change occurred with time. The observed changes are, in fact, the principal basis for believing that such a thing as evolution has occurred. About 2500 families of animals alone are identifiable in fossils. The average time for their existence was about 75 million years, and roughly a third of these families remain extant. Despite extinctions, the number of animal families has increased since the Cambrian. Something less than 100 families during the Cambrian itself increased to about 300 in the Silurian and Devonian. This number remained more or less constant until the Triassic, when there was a decline to about 250 families as a result of mass extinctions of many marine animals during the Permian; but afterwards there was a nearly linear increase in the number of families up to the present number of nearly 900. The much-talked-about extinction of the dinosaurs in the Cretaceous was more than balanced by the evolution of other families.

With respect to animals, they were on the earth in primitive form at least as early as the late Precambrian in what is known as the Ediacarian and related fauna. The name is derived from their most spectacular example in deposits of tidal flats off South Australia, but fossils of the same general age have been found in southeastern Newfoundland, England, the Soviet Union, and most recently North Carolina. (For a detailed discussion and key to the literature, see Cloud *et al.,* 1976.) In the Carolina case, the name *Vermiforma antigua* has been given to the metazoans that lived in deep water on a gentle marine slope near what is today Durham, N.C. Their visible traces in laminated, greenish, volcaniclastic sedimentary rock indicate they were large (up to one meter long), soft-bodied, wormlike animals related to the Annelida. The living worms were probably detritus feeders. The fossils were dated (zircon Pb–U) from nearby rocks to be 620 ± 20 million years old (Glover and Sinha, 1973). They are thus about the same age as the Newfoundland fauna, estimated as being 610–630 million years old (Anderson, 1972). The English and Soviet examples have secure dates a bit earlier, being nearly 680 million years old. There may also have been jellyfish and sponge spicules by this time, but after the time of the Ediacarian fauna, about 600 million years ago, the well-known and enigmatic explosion of life occurred, as seen in the fossil record.

Cambrian rocks of America alone have yielded more than 1200 species of animal. They were typically calcareous, shelly invertebrates such as the brachiopods (bivalves similar but not identical to clams) and gastropods (snails). The dominant early form was the famous trilobite (related to the extant ostracod), which attained a size of 18 inches in *Paradoxides,* the largest animal alive at the time. However, trilobites were not the first invertebrates of the period. The very earliest portion of the Cambrian (constituted by the Tommotian strata, the base of which reflects the initial appearance of an abundance of animals with calcareous skeletons) was characterized by archaeocyaths, molluscs, hyoliths, algae, and "problematica." The Tommotian molluscs were small or minute limpet-shaped, planispiral, or helically coiled univalves. (For a discussion of them in depth and for the probabilities of how they relate evolutionarily to subsequent taxa, see Runnegar and Pojeta, 1974.) Following the brachiopods, gastropods, and trilobites late in the Cambrian came the cephalopods (chambered shells) and foraminifers (protozoa), followed by starfish (and other echinoderms) and corals (coelenterata) in the Ordovician. By the middle Ordovician the seas also had vertebrates, evident in fish; and land invertebrates were represented by millipedes, scorpions, and spiders. Subsequently we find fossils of land vertebrates in the Devonian amphibian stegocephialins, to be followed still later in the Pennsylvanian by *Eosaurus,* the first reptile, which strangely seems to have coevolved with the first land snails. The vertebrate scale is then ascended to the mammalian stage with a few fossils in the late Triassic at the beginning of the Mesozoic. Their subsequent changes become clearer and clearer with increasing abundance of fossils as the age of the sediments becomes younger and younger, and mammalian evolution in the late Cretaceous but mostly in the Cenozoic to primates and finally to man is especially well documented as time-dependent in fossils. Depending on what we shall call man, he himself appeared only 1 to 2 million years ago and then underwent remarkably fast further evolution to his present state.

Algae were present in the Precambrian (see below), but not until nearly 200 million years after the beginning of the Cambrian itself did the record reveal vascular plants (tracheophytes) on land. The oldest fossils of the usual kind (visible to the unaided eye sometimes called megafossils) were first found in Scotland dating in the Lower Devonian. Predevonian remains have, however, been found in marine strata. For instance, recent analysis of fossil spores (microfossils) appearing to be those of land plants (Gray *et al.,* 1974) has shown their presence in a Silurian marine stratal sequence on the island of Gotland in the Baltic Sea. Moreover, it was shown that the abundance of spores is a function of water depth and shoreline proximity, suggesting their origin in land plants. In any event, by the close of the Devonian, moderately advanced plants [all evergreen, e.g., ferns, seed ferns (primitive embryophytes), scouring rushes, scale trees, and precursors (cordaites) of the conifers] were present forming forests. During the subsequent carboniferous periods great assemblages of these plants developed, leading in the Pennsylvanian to the great coal-forming swamps. The vascular plants occurring then are known from studies of petrified peat ("coal balls") and more recently from the study of spores and pollen (palynology). Five major groups of plants were in these swamps: lycopods, ferns, pteridosperms, cordaites, and sphenopsids. It has been possible to show from palynology that broad climatic shifts at the Desmoinesian–Missourian (Westphalian–Stephanian) boundary, which was probably multicontinental in scope, resulted in a shift from a lycopod-dominated flora to one in which tree ferns were the major element (Phillips *et al.,* 1974). As with the animals, which evolved from egg layers to placental mammals to menstruating primates, invasion of the land by plants led to great changes associated with reproduction, which took place in a time matrix so far as fossils currently document what happened. Conifers (gymnosperms) appeared about the middle of the Pennsylvanian, but flowering plants (angiosperms) were clearly present not for another 200 million years. Undoubtedly, angiosperm pollen first appeared in the geological record during the Albian or uppermost stage of the Lower Cretaceous (Brenner, 1976). Well-preserved flowers, which are rare, have only recently been obtained from Eocene formations in southeastern North America that are 50 million years younger than the earliest pollen grains (Crepet *et al.,* 1974). Nevertheless, the flowers have allowed the interesting conclusion that the degree of evolution of pollen paralleled that of floral parts. Much interest has in fact been attached recently to palynology. Thus, Walker and Skvarla (1975) have presented an evolutionary scheme for vascular plants that depends on the exine structure of the spore pollen. Taxa in the Magnoliaceae, Degeneriaceae, Eupomatiaceae, and Annonaceae (all families belonging to the order

Magnoliales) are on this basis quite primitive. Interestingly, aspects of the fossil record, present distribution, and various flower characters also indicate Magnoliales to be the earliest or at least among the earlier of angiosperms. The order is represented today by the "tulip" (*Liriodendron tulipifera*) and *Magnolia* trees in the southeastern region of the United States.

The current existence of primitive angiosperms has its parallel in many other plants. Thus, the primitive gymnosperm *Ginkgo biloba* survives today as a representative of the Ginkgoales flourishing in the middle Mesozoic but present as early as the late Permian as seen from leaf fossils. A widely distributed Jurassic species with leaves similar in size and shape to *G. biloba* was *G. digitata,* found in Oregon and Alaska as well as Australia, Japan, Turkestan, and England. *G. adiantoides* from the Eocene, found in various places in N. America, may actually be identical to *G. biloba.* Another "living fossil" among the gymnosperms, though not as old as the ginkgo, is constituted by the nine extant genera of the Cycadales, which have their roots in the Jurassic period. We also have with us the older ferns, such as the Osmundaceae, not to mention the very ancient algae. A similar situation occurs among animals.

The exquisite chambered nautilus (a cephalopod) survives in the Pacific Ocean from the Philippine to the Fiji Islands at depths up to 2000 feet in essentially the same form as its ancestors of 180 million years ago. Even more dramatic is the coelacanth, a Crossopterygian fish, first discovered in the living state in a specimen weighing 127 lb caught in 1938 by a trawler at 40 fathoms off East London on the African west coast. It was brought to the attention of the scientific world by M. Courtenay-Latimer, curator of the East London Museum, who communicated the information to Dr. J. L. B. Smith of Rhodes University in Grahamstown. In 1952 another specimen was caught near an island in the French Comoro archipelago north of the Mozambique Channel, and by 1955, with fishermen alerted to interest in the fish, nine new specimens 4–5 ft long and 5–20 years old became available in good condition for study. The fish, which has a skull similar to primitive amphibians but is viviparous (Smith *et al.*, 1975), was given the name *Latimeria* (for Ms. Latimer) *chalumnae.* It is remarkable not only for its anatomical and reproductive aspects, but because except for being a bit larger than its ancestors it remains essentially unchanged from fossil coelacanths that lived in the Devonian up to 400 million years ago. Similarly, the extant horseshoe crab, *Limulus polyphemus,* except again for being (five times) larger than its ancestors, is the same as fossil representatives 300 million years old. Other "living fossils" include the Mesozoic nut clam (*Nucula*), a brachiopod (*Lingula*), and a Mesozoic lizardlike reptile (*Sphenodon*). There is in addition the frog line. The earliest frog fossils are from the Jurassic, and except for having ribs they are not very different morphologically from living frogs of the families Ascaphidae and Discoglossidae.

Although the fossil record does indeed document change in time, the existence of "living fossils" poses some very difficult theoretical problems as to why they did not change as many other species did. The "living fossils" mentioned are actually only representative of a much more general phenomenon exceedingly well documented at higher taxonomic levels, namely, that all of the major early forms of life still exist. We still have bacteria, algae, worms, mosses, fish, molluscs, lycopodia, ferns, gymnosperms, amphibians, reptiles, etc. In fact, few forms of life at the higher taxonomic levels that ever existed according to the fossil record do not exist today. The extinction of some particular adaptation—for instance, of the arthropods, say, a trilobite; or of the reptile group, say, a dinosaur; or of the feline group, say, a saber-toothed tiger—is trivial compared to what is not extinct. Moreoever, there is no evidence in fossils for real precursors of the major groups. They simply appear out of nowhere in the geological record. There are no "missing links" of consequence. What the fossil record does show and what has been written about in detail and in many places is that within a very narrow taxonomic group, say, horses starting with the famous "dawn horse" (*Eohippus*), the fossil record demonstrates change in relatively small (and therefore interesting) anatomical increments from the past to the present. Further paleontological work, especially with pollens and spores, will probably expand our knowledge of such microchanges, but we have with us something more profound than the question of the origin of species.

Species, if indeed there is any such thing in a fundamental sense, seem to have appeared as an adaptation. Current thought is that this adaptation was probably ecological, almost Lammarckian in character. The biochemical mechanisms, however, remain obscure. Still more obscure are the origins even anatomically of the higher taxons. Where did the Ediacarian fauna come from? Do they really relate in a familial sense to the Cambrian explosion of arthropods? Where did vertebrates come from? What is the origin of the fungi? Did the algae really lead through mutation and a direct reproductive line to the mosses, and did they really lead to the ferns, etc.? If so, why does the fossil record show only the appearance of algae, mosses, ferns, etc., and no intermediate forms? How many lines of evolution are there anyway? Only one? Two? A thousand? The fossil record does not give us an answer to these questions any more than it did to those of Darwin's time. Yet, it is through paleontology that we have recently been afforded some exciting partial answers. The great silence of the Precambrian eons is in the process of being broken, and we can now glimpse life at its very beginnings.

In 1940, limestone ($CaCO_3$) formations were discovered in the Transvaal of Africa. These formations and subsequently discovered ones like them (dolomitic ones, which includes $MgCO_3$, and siliceous ones) all called stromatolites are laminated sedimentary structures caused by the trapping and binding of sedimentary particles by algae or photosynthetic bacteria or both, as discussed in greater detail by Nagy (1974) and Brock (1978). The ancient stromatolites are usually found on "old shields," i.e., Precambrian rocks not covered by other sedimentary structures. Stromatolites, however, are currently being formed, as in the case of Shark Bay, Australia (Playford and Cockbain, 1969), and in hot springs in Yellowstone Park (Brock, 1978). In the latter case Brock (1978) has recently discussed at some length a series of investigations he and his students made over several years. Although there may be a wide variety of stromatolitic types in Yellowstone, Brock concentrated on two types both of which are siliceous. Opaline silica, precipitated by cooling and evaporation of the water, encrusts the filaments of the microorganisms and slowly builds a permanent rock. In the first type, conical-shaped structures are formed at 32–59° by a filamentous blue-green alga dominated by *Phormidium tenue* var. *granuliferum*. These stromatolites are related structurally to the Precambrian ones called Conophytons. Below 32° coarsely filamentous blue-green algae, e.g., *Calothrix coriacea,* dominate the stromatolitic mats, which become stratiform or nodular. The second type of structure has the form of flat, laminated mats produced by photosynthetic bacteria in the genus *Chloroflexus* in conjunction with blue-green algae in the genus *Synechococcus*. If the light intensity is reduced or the sulfide concentration increased, blue-green algal development is suppressed with maintenance of mat formation. In high-sulfide springs pure *Chloroflexus* cultures exist. Brock (1978) takes this to mean that stromatolitic formation can be achieved either by the bacterium alone or by populations dominated by blue-green algae.

Generally, time erodes stromatolites, but in some cases protection for one reason or another has led to ancient structures that in some cases preserve fossils of the organisms that formed them. The oldest stromatolite is in Bulawaya, Southern Rhodesia. It has been dated as being 2.7–3.1 billion years old. Although a morphological imprint of organisms has not been observed in it, the depositional character of the rock closely resembles ones known to be microorganismically derived. Even when imprints can be found, most of the ancient stromatolites unfortunately yield fossils that are poorly preserved as a result of diagenetic recrystallization of the carbonates. The so-called bedded cherts (a type of flint with a silicate base), however, have yielded good fossils. The cherts often also have interstitial laminae of iron. The Australian Bitter Springs deposit dated (Rb–Sr) at about 900 million years old is of this sort. Fossils of organisms that resemble blue-green algae of the genera *Oscillatoria* and *Nostoc* have been found by Schopf and Barghoorn (1969) in the Bitter Springs deposit. Analysis of the fossils of this deposit also were thought to have revealed 20 types of blue-green algae and three species of bacteria. It was even thought that eukaryotic green algae, named *Glenobotrydion aenigmatis,* might have been present going through mitotic division (Barghoorn and Schopf, 1965). The "green algae" and the "mitotic division" were unfortunately subsequently shown to be artifactual (Knoll and Barghoorn, 1975), and the latter work on artifacts suggested that the diversity in the cyanophytes was probably also wrong. Knoll and Barghoorn (1975) now feel it likely that cells were present but representing only a single species of blue-green alga (*Chroococcus*) in the Bitter Springs deposit. Eukaryotic cells were also reported by Cloud and his associates in the Beck Springs (1.3 billion years old) formation in California (Cloud *et al.*, 1969). In view of the work on artifacts by Knoll and Barghoorn (1975), however, the authenticity of these eukaryots is brought similarly into question. Nevertheless, during the last 15 years 30 well-preserved microfossil assemblages from the Precambrian have been reported. (For a key to the literature, see Nagy, 1974; and Schopf, 1974, 1975.) In North America most occur in stromatolitic cherts and represent shallow-water benthic mat communities. Carbonaceous shales from the Soviet Union, Europe, and America have also revealed abundant and well-preserved filamentous and spheroidal microfossils. An example is the Middle Proterozoic Newland Limestone exposed along a road-cut of U.S. Highway 89 in the Little Belt Mountains just southeast of Neihart, Montana. This deposit was dated at 1.4 billion years old. It contains fossils of at least four species of filamentous blue-green algae similar to *Nostoc* as well as several spheroidal planktonic forms believed tentatively to represent the encystment stage of eukaryotic algae (Horodyski and Bloeser, 1978). This is the oldest North American shale known to contain fossils of cells. Other geological deposits that have yielded microfossils are the Gunflint chert of Canada (unicellular and colonial filamentous cells, 2 billion years old) (Barghoorn and Tyler, 1965; Cloud, 1965), the Transvaal Sequence in South Africa (2.2 billion years old) (Nagy, 1974), and the Fig Tree Series in eastern Transvaal, South Africa (rod-shaped bacteria, 3 billion years old). The filamentous organisms from the Gunflint are strikingly similar to the photosynthetic bacterium *Chloroflexus* (Brock, 1978). The microfossils of the Transvaal stromatolite are morphologically similar to the living blue-green genus *Raphidiopsis* (Nagy, 1974), although Brock (1978) feels unequivocal blue-green algal fossils are from younger deposits (1 billion years old). Thus, the present evidence from fossils would strongly suggest that cellular life may have been present as much as 3 billion years ago in the form of photosynthetic prokaryotes, viz., bacteria, with blue-green algae perhaps appearing later.

In closing this section on fossils we should like to mention that an attempt has been made to determine what type of carbon fixation was occurring in the distant past through measurements of the $^{13}C/^{12}C$ ratio in the Precambrian Onverwacht sediments in South Africa. It is unfortunately agreed that the data prove very little, if anything (Barghoorn *et al.*, 1974; Cloud, 1974). In addition, still more recent work shows that the $^{13}C/^{12}C$ ratios are related not necessarily only to fractionation by cellular photosynthesis; thermal cracking of organic matter, solar-proton "stripping," and alpha radiation from uranium can also alter the ratio substantially (Leventhal and Threlkeld, 1978). Moreover, the abundance of ^{13}C and ^{18}O in organisms has recently been shown to have complex ecological relationships (Emiliani *et al.*, 1978). The problem of ecology, as alluded to earlier in this section, just can no longer be ignored in assessing past biological phenomena. It has recently been shown, for instance, that an interpretation of the record of planktonic microfossils requires as much as anything else an understanding of the climatic conditions at the time the organisms lived; i.e., the species distribution depends on the characteristics of the sea's surface (Weyl, 1978). Actually, rather than analyzing evolution from the species distribution, one can more surely analyze climate by assuming the older species had the same characteristics as the extant ones.

The fossil record is no doubt a fact showing change in time, but as with all facts it is opening more questions than it answers. What really were the cells seen in the ancient fossils? Ancient molecular fossils of phytol suggest they may indeed have been photosynthetic, but were they all photosynthetic? If so, where did the nonphotosynthetic line come from? Did it have a separate origin, or did algae lead to mushrooms and man? Were there originally many species of blue-green algae, or only one which mutated in response to ecological pressures? Or did they or any other cell-line mutate for other reasons? What kind of reasons might there be? Were nonphotosynthetic bacteria present originally? If not, when and how did they arise?

C. EVOLUTIONARY LINES

It will be appreciated from Section B that the record of the rocks remains enigmatic on many essential questions. Despite this, anatomical, chromosomal, biochemical, and other relationships among various extinct and extant life forms, coupled with what fossils do tell us, have led to the postulation of a variety of evolutionary lines or "trees." The purpose of this section is to present some examples of such lineages to illustrate current thinking. It will be evident from what follows that multiple lines of evolution, at least after the Cambrian, are generally assumed. The validity of the "trees" will be assessed in part in subsequent sections dealing with lipid biochemistry.

Among the oldest, clearest, and most abundant of the Cambrian fauna were animals in the phylum Mollusca, with seven classes now alive: Aplacophora, Polyplacophora, Monoplacophora, Gastropoda, Cephalopoda, Pelecypoda, and Scaphopoda. Stasek (1972) has suggested that these extant classes represent three separate lines (three subphyla: Aplacophora, Polyplacophora, and Monoplacophora) with no known intermediate forms bridging "the enormous gaps between any two of the three lineages." Runnegar and Pojeta (1974) view molluscan evolution somewhat differently. They believe the fossil record indicates only two major lines: (1) the Aplacophora and (2) the Monoplacophora. The latter are believed to be the precursors of the Gastropoda, the Cephalopoda, possibly the Polyplacophora, and through the Rostroconchia also of the Pelecypoda and Scaphopoda. Most of the earliest of the molluscs (present in the Tommotian or oldest strata of the Russian Cambrian) seem to have been tiny monoplacophorans classified in the superfamily Helcionellaceae. It is these that Runnegar and Pojeta (1974) consider to be the precursors to all extant molluscan classes except the Aplacophora. Both groups of investigators believe the Aplacophora had their own Precambrian origin not now assessable from fossils. The other major group of animals in the Cambrian fossils are the Trilobita, which are first found in the penultimate Russian Cambrian stratum (Atdabanian). In America, fossils believed to be molluscan have also been found below the oldest trilobite fossils (Taylor, 1966). The trilobites therefore appear to have evolved later than the molluscs. While the former do not seem to have given rise to other major taxons up to the late Ordovician, the seven classes of mollusc had already formed by this time from the two, three, or whatever number of precursors they had.

A wider view of animal evolution has been given by Valentine and Campbell (1975). These authors break animals into the Parazoa, Diploblastica, Early Coelomata, Later Coelomata, Aquatic Vertebrata, and Tetrapoda. Each of these groups is thought to have had its origins in the Precambrian Protista. The Parazoa, e.g., Porifera, are thought to have possibly diverged from the Protista in the late Precambrian about 700 million years ago. All the others are speculatively given lineages going back much further to Protista in the early Precambrian. The Diploblastica (Coelenterata and Ctenophora), Triploblastica (Platyhelminthes, etc.), and Early Coelomata (Annelida, etc.) are supposed to have formed in the late Precambrian between 700 and 800 million years ago. About the beginning of the Cambrian, the Later Coelomata (molluscs, arthropods, brachiopods, etc.) are then presumed to have developed from the Early Coelomata; for example, the annelids are thought to have given rise to the arthropods at this time. Subsequently, in the Ordovician, about 450 million years ago, the Later Coelomata themselves presumably branched, giving the Aquatic Vertebrata, among which were the lung fishes, which in turn about 350 million years ago may have given rise to the amphibians. Then, at the base of the amphibian line about 300 million years ago the reptiles are thought to have developed from the amphibian line, and, still later just before and after the 200-million-year mark, respectively, the mammalian and avian lines are believed to have evolved from the reptiles.

With respect to very early evolution, Valentine and Campbell (1975) suggest that animals with a coelom (body cavity) may have originated in primitive coelomates that had evolved from seriated flatworms in the late Precambrian about 700 million years ago. These early Coelomates then may have given rise, perhaps 650 million years ago, to four main stocks. One of these, seriated pseudometamerous organisms, are supposed to have been the progenitors of the Polyplacophora, Bivalvia, Monoplacophora, Cephalopoda, and Gastropoda. A second, the oligomerous lophophorates, are presumed to be the ancestors of the Ectoprocta, Phoronida, Inarticulata, and Articulata, while the third, the oligomerous deuterostomes, may have led to the Echinodermata, Hemichordata, Urochordata, and Chordata, the latter of course including vertebrates. Finally, these authors suggest that the Annelida, the Uniramia, and the Trilobita–Crustacea may have originated in the fourth (a metamerous organism) of the main stocks derived from the primitive coelomate. While not hazarding a guess as to the origin of the aboriginal flatworms, Valentine and Campbell (1975) suggest that the key to the change observed in the fossil record rests on an evolution of the regulatory portion of the genome that resulted in new anatomical plans. Once these new plans were established, rapid evolutionary radiation occurred through, at least in part, ecological phenomena as described by models such as those of Simpson (1953).

The phylogeny of seed plants, which first appeared in the fossil record in the Silurian and Devonian, has been viewed variously in detail by various authors. However, many believe gymnosperms gave rise to the more advanced groups. An example of such a lineage, taken from the work of Ehrendorfer (1971) and using his nomenclature, is: Psilophytes to extinct Progymnosperms in the Devonian, which then gave rise to extant gymnosperms on the one hand and angiosperms on the other. The ginkgos, cycads, Pinidae, and Taxidae are viewed as constituting four distinct gymnosperm lines. The latter two (Pinidae and Taxidae) are thought to have originated in the Permian and early Triassic, respectively, through unknown Carbonaceous and Permian intermediates evolving from the cordaites, which then became extinct at the end of the Permian. The ginkgos are viewed as having a separate origin in the progymnosperms, but the cycads are thought to have arisen through the Lyginopteridatae, which then became extinct at the end of the Jurassic. The Lyginopteridatae are also considered to be reasonable precursors of angiosperms (appearing in the late Jurassic) through unknown intermediate forms evolving in the Permian just after the carboniferous period. The earliest of angiosperms (Magnoliidae) are believed to have subsequently evolved into the monocots (Liliatae) and dicots (Magnoliatae). This gives us, then, three

extant supergroups: the coniferophytes, the cycadophytes, and the mono-
cotyledenous and dicotyledenous magnoliophytes. Except for the cycads,
which have very limited current representation, these plants are believed to
have undergone extensive ecological–adaptive evolution (Doyle and Hickey,
1976), leading to the large array of flowering plants present today. Although
the detailed phylogenetics even of the more recent plants is quite speculative,
the interrelationships and possible phylogenetics of tracheophytes have been
discussed by Cronquist (1968) and Hutchinson (1969).

Lower on the tracheophyte scale we encounter the many varieties of fern
that may have originated in the Devonian and were certainly well established
in the Carboniferous in some of their present forms. So much controversy still
surrounds just the classification of fossil and extant ferns (see Jermy *et al.,*
1973, for a discussion and key to the literature) that little progress has been
made in the examination of their evolution. Various lines of ascent have been
suggested for 100 years, but the uncertainties are so great that we prefer not to
present any of the hypotheses.

Still further down the scale with respect to both organization and time are
the algae. Based on the work with fossils in the Bitter Springs Formation,
which has serious problems (Section B), some investigators have felt that the
oldest aglae may be the coccoid Chlorophyceae, (Ulotrichales and Chlorococcales). Younger and more secure fossils (just before the Cambrian) are
thought to represent Dasycladales, calcareous Chlorophyceae, and algae in
the family Botryococcaceae or the order Chlorococcales. The Cambrian
reveals Siphonales, and later in Jurassic sediments one finds a more diverse
assemblage, including all the above-mentioned algae as well as the appearance
of the Volvocales, Zygnematales, Oedogoniales (going back actually to the
Devonian), the Cladophorales and Siphonocladales. Ordovician, Silurian,
Devonian, and Mississippian sediments have many well-preserved fossils.
Among these, the most frequently encountered and best-preserved fossils are
those of the calcareous organisms (Codiales and Dasycladales of the
Chlorophycophyta and Solenoporaceae and Corallinaceae of the Rhodophycophyta). *Phyco* (Greek, "seaweed"), incidentally, is a combining form in the
division classification some botanists use to emphasize the difference between
eukaryotic algae and other forms of life, e.g., the prokaryotic Cyanochloronta
(Bold and Wynne, 1978; Papenfuss, 1946). The fossil record suggests that the
common extant seaweed groupings, Rhodophycophyta and Phaeophy-
cophyta, were present as early as the Cambrian. The Charophyta (stone-
worts), which are not phycophyta, seem to have appeared in the Silurian. The
Chrysophycophyta (Bacillariophyceae, Chrysophyceae, and Xanthophyceae)
and the Euglenophycophyta appeared in the Cretaceous. Fossils of the
diatoms (family Bacillariophyceae) are well represented from the Jurassic
continuously to the present. Appearing in several deposits are microalgae,

morphology, classification may often have little to do with phylogenesis in the sense of analyzing precise progenitors. Two anatomically similar forms of life may actually have had different rather than the similar origins which the morphology would suggest. A specific example of such parallel evolution can be taken from the frogs (order Anura, 150 million years old from the fossil record). The extant Hylidae family of tree frog is thought to have arisen in the New World from a bufonoid stock, while a ranoid stock is believed to have led in the Old World to the Rhacophoridae family. Both of these families are so closely related anatomically that assignment of a particular species to one or another of them can be quite difficult without geographical information.

Independent evolution of anatomical and biochemical parameters is also well documented, and again the tree frog will serve for illustration. In the New World genus *Hyla,* the two species *H. eximia* and *H. regilla* are almost the same anatomically but definitely different from species in another genus, e.g., *Acris crepitans.* Despite the morphological similarities of *H. eximia* and *H. regilla,* the albumins of these two animals are quite different, as different in fact as the difference in the albumins of two species in different genera, e.g., *H. eximia* and *A. crepitans.* If, as done by Maxon and Wilson (1974), one takes the protein (albumin) structure as a marker of phylogenetics, then the anatomical structure must not be a marker for lineage. The differences in the proteins of the two *Hyla* species would indicate two different detailed lines of evolution, and the similarities in anatomy would then have to represent an anatomical convergence of the two lines. One can, of course, argue the reverse. Perhaps the similarities in anatomy represent a common lineage. This would require that divergence in evolution of the protein had occurred. Whichever way the facts are approached it becomes clear that the anatomy and the albumin of these frogs did not evolve hand-in-hand. Interestingly, the two *Hyla* species are so alike morphologically that a third type of *Hyla* (the so-called "wrightorum" frog) exists that is anatomically intermediate and has been classified by some investigators as a subspecies of *H. eximia* and by others as a subspecies of *H. regilla.* Maxon and Wilson (1974) examined the albumin of the wrightorum frog and found it nearly identical to that of *H. eximia,* which, it will be recalled, is very different from that of *H. regilla.* Conclusions about the wrightorum variety depend on whether we wish to describe the frog in terms of what it is and what it can do or in terms of where it came from (which entails a choice of phylogenetic markers). The various descriptions seem to us each to be valid, but they may not always be identical. The wrightorum frog might well be a morphological subspecies of *H. regilla* while being a phylogenetic subspecies of *H. eximia.* This general intellectual problem has been well documented at detailed structural and biosynthetic levels with the lipids. Exactly the same compound may arise in two different organisms but by different routes. Should we use the route or the compound as a phylogenetic marker? The problem can be broadened. Suppose two compounds, each in a different organism, have not the same but different structures and yet arise by the same general route with the addition or deletion of some particular step. Should we use the structure or the route to examine phylogenetics? These are among the questions we shall consider at some length in the rest of the book.

CHAPTER 5

The Origin of Oxygen

A. GEOLOGICAL AND ASTROPHYSICAL EVIDENCE

Several different kinds of things having to do with oxygen have led to a great deal of speculation on the role of this substance in evolution. We will first summarize the current status of the arguments and then in Section B examine how our knowledge of lipids bears on the problem.

Early in the second quarter of the present century the sun was found to be mostly (87%) hydrogen (Russell, 1929), and with more work the atmospheres of the outer planets, e.g., Jupiter, were shown to have an abundance of hydrogen, helium, methane and ammonia. This led many people, sparked by the speculations of Oparin (1924, 1953) and Urey (1952), to "conclude" that the primitive atmosphere must have been in a highly reduced state with N, O, and C existing as hydrogenated derivatives (NH_3, etc.). Actually, the origin of the planetary atmospheres (and to a lesser extent the planets themselves) is far from understood. There are roughly two types of planets, the inner ones (Mercury, Venus, Earth, and Mars in increasing distance from the sun) and the outer ones beginning with Jupiter. In between these groups is the asteroid belt, containing objects as large as 600 miles in diameter. It is now known that the atmospheres of the inner planets are not similar to each other and that they are quite different from those of the outer planets. A most striking difference between our own oxygen-rich atmosphere and the highly reduced atmosphere of Jupiter is found on our neighbor, Venus. Exacting values for the composition of the Venusian atmosphere (pressure 91 atm) were obtained by gas chromatography on board the Pioneer space vehicle on December 9, 1978 (Oyama *et al.*, 1979). As seen from Table 12, the atmosphere is primarily constituted by carbon in its most highly oxidized state (CO_2), which causes a greenhouse effect bringing the surface temperature to 455°C. Current astrophysical views are that about 4 billion years ago something swept the

Table 12. The Atmosphere of Venus[a]

Substance	Contribution
Carbon dioxide	96.4%
Nitrogen as N_2	3.41%
Water vapor	0.14%
Oxygen as O_2	69 ppm
Argon	19 ppm
Neon	4.3 ppm
Sulfur dioxide	186 ppm

[a] From Oyama et al. (1979).

inner planets bare of their primal atmospheres. One candidate is the so-called "T tauri wind," characteristic of newly born stars. The principal reason for believing that the modern atmospheres are not primal but have evolved during the interim time is the lack of correlation, beginning with the work of Brown (1949), between the compositions of planetary noble gases and those of the sun. (For deeper discussions and a key to the literature, see Waldrop, 1979, and various articles in the February 23, 1979, issue of *Science*, which is devoted to Venus.)

There have been other, not terribly convincing, reasons for believing in a reduced atmosphere on earth consisting of the following major points as discussed by Lemmon (1970): (1) the presence in meteorites of carbon, carbides, hydrocarbons, metallic or ferrous iron, and phosphides; (2) the conversion of methane, ammonia, and water to amino acids and other biochemicals in the laboratory; (3) the "deleterious effect" of oxygen "on many aspects of cell metabolism"; and (4) the discovery of ferrous iron in some Precambrian sediments. Except for the last of these it will be obvious that none constitutes direct information. Furthermore, the hydrocarbons in meteorites are by no means surely of extraterrestrial origin (Chapter 4, Section A), and the supposition of the deleterious effect of oxygen on cells is not biologically tenable. Many bacteria usually considered quite primitive operate aerobically. In terms of direct evidence, that leaves us with the ferrous-iron sediments, which are well known to need much more definition. It is perhaps also worth mentioning that other inconsistencies exist in the older views. Qualitatively, if methane and ammonia with molecular weights of 16 and 17, respectively, distilled off the earth due to their light weight, why would most of the xenon, with an atomic weight of 131, leave? The noble gases in our atmosphere actually have values considerably less than their cosmic abundance. And why should ammonia and methane leave if water (mol. wt. 18) did not? Perhaps H bonding is the answer. Finally, as J. Friend pointed out to one of us, why didn't ammonia dissolve in the water, leaving us with an alkaline ocean? Answers to these and other crucial questions about our atmosphere are simply not available. We just do not know where or when our atmosphere came from, especially in terms of N_2, O_2, H_2O, and CO_2. Since the last three are biochemically critical substances, we are left with a very important mystery about our beginnings, but some fashionable speculations have been put forward as discussed in what follows.

The most common hypothesis is that the atmosphere is a secondary one having resulted from volcanic emissions or other outgassings. The primary atmosphere mimicking cosmic abundance was either lost very early or, as supposed by Berkner and Marshall (1965), never existed. Since the effluents from current volcanoes do not contain appreciable amounts of molecular oxygen, the hypothesis invokes solar radiation in photolysis of water ($2H_2O \longrightarrow 2H_2 + O_2$) as the source of oxygen. (For detailed discussions and a key to the literature, see Berkner and Marshall, 1965, and Cloud, 1968.)

Despite the manifest uncertainties, we do know that ozone absorbs ultraviolet radiation from the sun and that there would be a lethal effect of such (unabsorbed) radiation on living cells. Without oxygen, there presumably could be no ozone. Without ozone, there presumably could be no cells. So, since life was present on earth 3 billion years ago, there must already have been atmospheric oxygen to produce ozone. However, this argument is generally ignored or is explained away by still another hypothesis. Berkner and Marshall (1965), Sagan (1961), Hoering and Abelson (1961), and others have supposed that early life (blue-green algae) was protected by a habitat in shallow waters where the cells were less than 10 m deep. Presumably the blue-green algae were relegated to this narrow zone, because too much depth of water would prevent photosynthesis and too little would prevent UV screening.

Still another argument is current. It depends upon the existence of obligate anaerobic prokaryotes. The simultaneity of simplicity in organization and an anoxygenic metabolism leads to the supposition that these organisms are primitive forms with heterotrophic metabolism. In this connotation primitiveness implies a low degree of sophistication as well as primitiveness in time. Moreover, the implication of what an anaerobe does is presumed to reflect what anaerobes did in the early phases of the evolution of life. Even if the postulate of a secondary atmosphere were to supplant the Oparin–Urey model of a primary atmosphere, the existence of anaerobic prokaryotes is construed as evidence for the development rather than the prior existence of oxygen on the earth. The source of oxygen is taken to be the blue-green algae, the only prokaryotes known to possess photosystem II (PS II), allowing them to produce oxygen photosynthetically (Brock, 1973). With the appearance of the blue-green algae, the oxygen content of the atmosphere presumably increased with time until it reached a level high enough to support aerobic organisms. Since aerobic metabolism yields many times more energy per mole of substrate than does anaerobic glycolysis, much is supposed to have happened when a shift in the composition of the atmosphere took place. In particular, the newly developed oxygen is supposed to have arrived at a magic number of 1% of its present level at or just before the great biological diversification apparent in the Cambrian fossil record, thereby accounting for the "explosion" of life forms at this time. No consequential evidence beyond what has been cited here is available. Even accepting such ideas, we are left with a mystery about the role played by the anaerobic bacteria unless we suppose they were first, which the stromatolites' evidence may support (Brock, 1978) (Chapter 4, Section B-2), and gave rise to the blue-greens. However, there are phycological reasons, e. g., those phyletic relationships of photosynthetic prokaryotes containing either chlorophyll or bacteriochlorophyll and PS II or PS I and PS II (Fogg *et al.*, 1973), to believe that photosynthetic bacteria are not the progenitors of the blue-green algae (Bold and Wynne, 1978).

B. THE LIPID TESTIMONY

1. Anaerobic Biosynthesis

To get at the problem, we will have to make the tentative assumption that biochemical vestiges of the past might be found. Such an assumption is not unreasonable in view of the existence of anatomical analogs. Many aspects of embryology are good examples ("ontogeny recapitulates phylogeny"). Early in the ontogeny of a human being, the remnant of a tail is evident at the base of the spine. Again in the human being, a gill precedes the development of the lung, and more generally all vertebrate embryos appear morphologically similar very early in their ontogeny, suggesting a common origin (Huxley, 1969). The most profound vestige, of course, is the tadpole or fish stage of frog development, which is taken to document an aqueous origin of land animals.

At a biochemical level we would hope to find an anaerobic biosynthesis of lipids. It would be especially significant if we could show an anaerobic vestige in mature organisms as well as to show a shift from an anaerobic to an aerobic route during ontogeny. In view of the fact that the earth abounds in organisms that can be placed in an evolutionary hierarchy either on grounds of the paleontological record or on a morphological basis (which would include subcellular organization in this connotation), we have an added possibility. A cell that is morphologically and, if we are lucky, also paleontologically primitive might be found. This then should form lipids primarily or exclusively by an anaerobic route. As with all proofs, belief in a development of oxygen in the atmosphere would gain strength as more and more expectations are verified and exceptions are explained.

Our now extensive knowledge of lipids has shown that the early stages of biosynthesis actually are anaerobic, and it makes no difference what the cellular type is. Nor does it make any difference which of the two great lipid classes is involved. Both the fatty acid pathway and the isopentenoid pathway begin the same way, by the anaerobic enolate condensation of acetate to acetoacetate. In the fatty acid sequence (Majerus and Vagelos, 1967; Lynen, 1967; Bloch and Vance, 1977; Volpe and Vagelos, 1973), the enolate condensations continue on in a linear fashion to give an unbranched array of

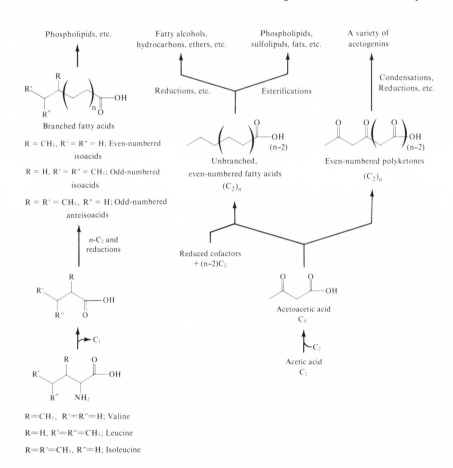

Figure 5. Anaerobic fatty acid pathway (CoA and ACP ester as well as malonate as a source of C_2 have been ignored).

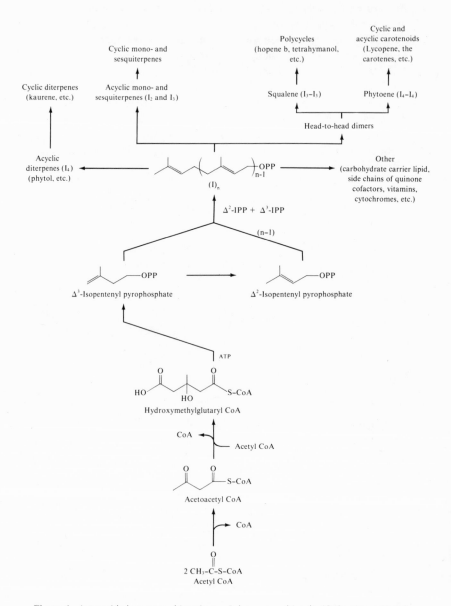

Figure 6. Anaerobic isopentenoid pathway. I, isopentenoid unit. (Only some examples of products are given. Stereochemistry of I_n can include *cis* double bonds.)

C atoms, while in the isopentenoid pathway (Nes and McKean, 1977; Nes, 1977; Goad and Goodwin, 1972; Goodwin, 1971; Bloch, 1965) the third C_2 condensation occurs on the keto group of acetoacetate (instead of on the carboxyl carbonyl), leading to a methyl branch. Further condensation of this isopentenoid unit by anaerobic electrophilic attack on π orbitals leads on to a branched array of C atoms. Terminally branched fatty acids also are formed when a deaminated branched amino acid (valine, leucine, or isoleucine) is substituted for acetate in the fatty acid sequence (Kaneda, 1977). Thousands of lipids are made in this way either directly or by the superimposition of other anaerobic processes such as further condensation, methylation from S-adenosylmethionine, hydrogenation of ethylenic or carbonyl double bonds by reduced pyridine and flavin nucleotides, dehydration, cyclization, rearrangement, and decarboxylation. A summary of this anaerobic system is given in Figures 5 and 6.

If all cells are derived from a primitive anaerobic condition, no better a vestige could seem to be found than the biosynthesis of lipids. Not only is an anaerobic route always used, but the products play vital roles. Some important examples of functional products fully anaerobic in their bio-synthesis are (1) many of the fatty acids (as esters) in membranes; (2) the isopentenoids, phytol and farnesol, in chlorophylls; (3) analogs of phytol in some membranes (e.g., in some thermophilic bacteria); (4) polycyclic products from the linear polymerization of acetate as antibiotics, e.g., methymycin; (5) isopentenoids either as such (terpenes) or as parts of another molecule (the cytokinins) as regulators of metabolism; (6) both isopentenoids, notably squalene, and linear acetogenins in the regulation of buoyancy; (7) lipids of both classes that regulate or protect against entry or loss of water; (8) unbranched lipids that act as a storage of carbon and energy; and (9) both branched and unbranched lipids functioning as hormones, pheromones, etc. Despite the beauty of the apparent correlation with the supposed anoxygenic beginnings, serious problems arise.

2. Anaerobic vs. Aerobic Desaturation of Fatty Acids

If membranes and other functional entities can be assembled with anaerobic lipids, can it be done in some better way aerobically? Do the more advanced cells supplant aerobic for anaerobic biosynthesis? The answer is unfortunately no. Lipids are always made basically by an anaerobic pathway. By basically is implied the process of assembling the carbon atoms into some aggregate, say, a linear C_{16} molecule, (e.g., palmitic acid), or a branched one with 30 C atoms, (e.g., squalene). However, an important detail seems to be just what we are looking for.

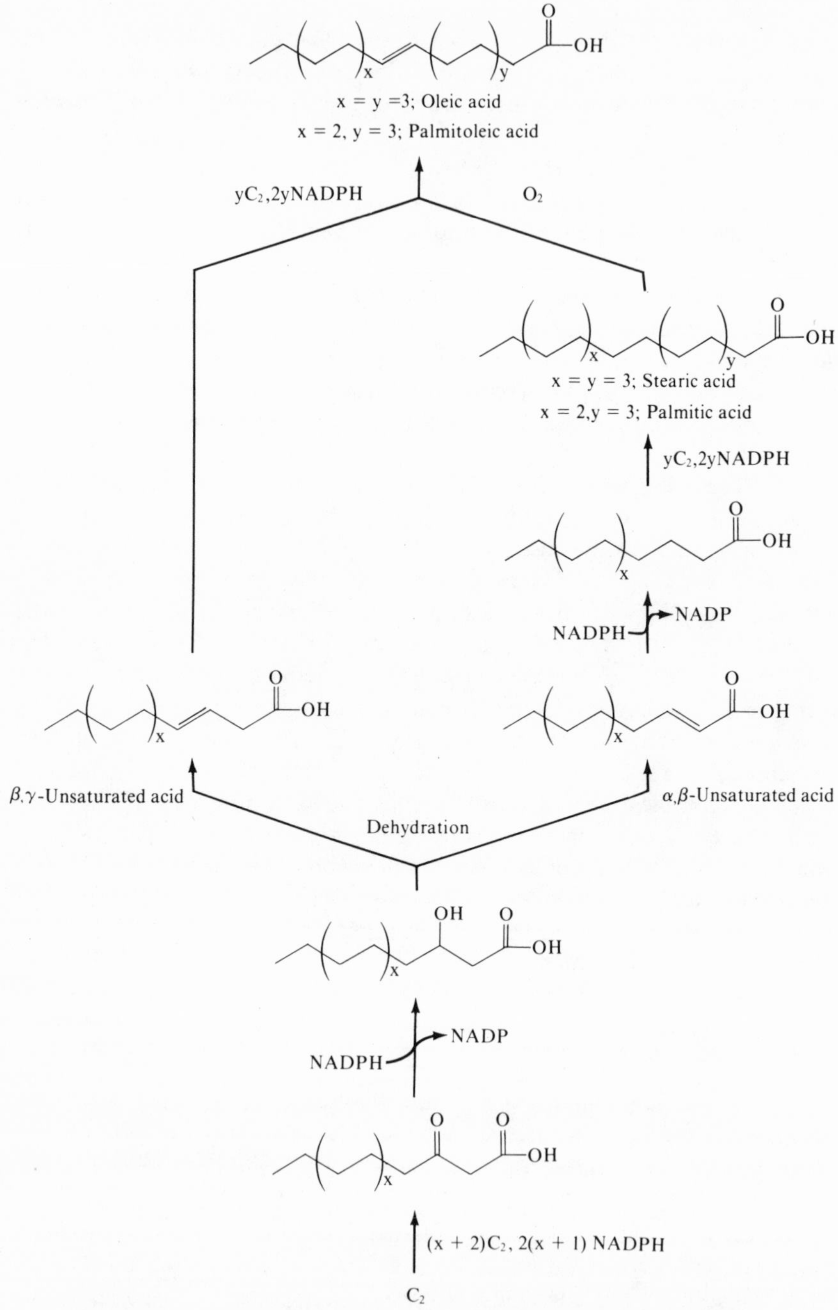

Figure 7. Biosynthesis of fatty acids with one double bond. (ACP Esters are ignored. The anaerobic route is to the left after dehydration, the aerobic route to the right.)

Table 13. Synthesis of Unsaturated Fatty Acids[a]
in Various Organisms

Aerobic desaturation	Anaerobic desaturation
Bacteria	Nonphotosynthetic bacteria
Pseudomonas aeruginosa	*Escherichia coli*
Mycobacterium phlei	*Clostridium kluyveri*
Corynebacterium diphteriae	*C. butyricum*
Various *Bacillus* sp.	*Treponema zuelzerae*
Yeast	*Sarcina* sp.
Saccharomyces cerevisiae	*Pseudomonas* sp.
Animals	Photosynthetic bacteria
Rat and rabbit liver	*Rhodopseudomonas spheroides*
Green algae	*R. capsulatum*
Chlorella vulgaris	*R. palustris*
Higher plants	*Rhodospirillum rubrum*
Spinacea oleracea	*Chlorobium limicola*
Glycine max	
Cucurbita pepo	
Epifagus virginiana	
Helianthus annus	
Sorghum bicolor	

[a]From Erwin (1973), Roughan (1975), Stumpf and Weber (1977), Nes *et al.* (1979a), Pugh and Kates (1979), Russell and Harwood (1979), and Gurr and Brawn (1970).

Two pathways to unsaturated fatty acids exist. One, operating in some bacteria (Bloch, 1969; Erwin and Bloch, 1964; Erwin, 1973) that in their prokaryotic nature are certainly primitive, is anaerobic (Figure 7); the other, operating more generally (Table 13), is aerobic (Figure 7) (Bloomfield and Bloch, 1960; Bernhard *et al.*, 1959; Hamberg, *et al.*, (1974). In the anaerobic route, suggested as primitive by Bloch (1969), the β-ketoacyl ACP derivative undergoes β,γ-dehydration at some stage of polymerization to give an olefinic double bond in the 3 position that does not suffer reduction as does the conjugated one in the 2 position derived from the usual α,β-dehydration. The Δ^3- and Δ^2-acyl ACP chains are then both elongated by additions of C_2 from malonate. When the Δ^3 bond is formed at the C_{10} stage, for instance, and three C_2 units are added, a Δ^9-C_{16} acid (palmitoleic) is produced. This double bond is needed to separate the molecule into two short chains of saturated C atoms in order to prevent a rigid intermolecular van der Waals' association and consequent loss of fluidity in the membrane. In the aerobic route the same result is achieved by direct O_2-dependent dehydrogenation of the saturated fatty acid, as in the conversion of palmitic to palmitoleic acid. Why, however, this is in some way "better" than the anaerobic route is not clear. It is not limited to eukaryotes (Table 13).

Escherichia coli forms unsaturated fatty acids by the anaerobic route either in the presence or absence of molecular oxygen. The failure of an extant prokaryote to shift from an anaerobic pathway in the absence of oxygen to an aerobic one when oxygen is present is both disappointing and paralleled in reverse by the failure of a eukaryote to shift backwards from the aerobic to the anaerobic route when oxygen is withheld. This cannot be demonstrated experimentally with many organisms, since the majority of eukaryotes, regardless of their degree of cellular differentiation, are obligate aerobes, but one is available, as discussed in the next subsection.

3. Anaerobic vs. Aerobic Growth of Yeast

The reason for an O_2 requirement of eukaryotes is often cited as a dependence on the operation of the cytochrome system of election transport for the derivation of energy. However, another reason is probably the dependence of the cell in detail on molecules derived from mixed-function oxygenase reactions. In at least one case (yeast) this is well examined. Yeast requires oxygen only for biosynthesis and then only for the formation of two sorts of substance, sterols and unsaturated fatty acids (Andreasen and Stier, 1953, 1954). In the latter case the O_2 requirement is for desaturation of fatty acids. The requirement in sterol synthesis is multiple: to epoxidize squalene, to remove the three methyl groups at C-4 and C-14, and to desaturate at C-5 and C-22. Under anaerobic conditions yeast makes only saturated fatty acids and squalene and will grow only if unsaturated fatty acid and an appropriate sterol are added.

The failure of yeast to revert to an anaerobic desaturation of the fatty acids in the absence of oxygen is a bit of a flaw in the fabric of our attempt to document an early anoxygenic atmosphere. In fact, the capability of yeast to operate at all in the absence of oxygen is a flaw. Anaerobic yeast cells do not seem to suffer from being anaerobic. They are very much alive and remain eukaryotic. In fact, they exhibit only slight differences from the cells grown aerobically. They are individually somewhat smaller and grow to a slightly smaller population at stationary phase, and their mitochondria disappear at least in the form observable (by electron microscopy) in aerobic cells. It is by no means obvious what if anything aerobic metabolism does energetically for the cells. On the other hand, there is a clear ecological advantage in terms of biosynthesis. Aerobic cells become independent of an exogenous source of unsaturated fatty acid and sterol. At the same time, by having a restricted requirement for products of aerobic metabolism the organism can live in an anaerobic environment that provides these substances, e.g., at the bottom of a pool. Such an environment, however, must be very sophisticated, because it must provide unsaturated fatty acids and sterols that can be formed only in the presence of oxygen. This means that the environment, though lacking oxygen at the moment, must be a product of an environment having oxygen. It therefore seems clear that the ability to shift from an aerobic to an anaerobic mode cannot be considered primitive or a vestige of an anoxygenic environment. Quite to the contrary, yeast must represent a highly evolved cellular adaptation (specialization) to existing conditions that allows it to flourish in a particular niche. There is really no more reason to assign primitiveness to anaerobic yeast than to muscle cells, which also utilize anaerobic glycolysis. Were the shift to an anaerobic mode a reversion to a more primitive condition, we would have hoped the yeast would exhibit other vestiges, which it does not appear to do. One possibility is the use of non sterols (or no sterols) in its membranes. We (Nes and Sekula, unpublished observations) have found that tetrahymanol, which is derived biosynthetically by an anaerobic process and functions in the place of sterols of certain protozoan membranes (see below), will not replace sterols in anaerobic yeast. We have also shown that the sterol requirement of anaerobic yeast is highly sophisticated and tied very precisely to aerobic metabolism (Nes *et al.*, 1976a, 1978a). In particular, neither squalene nor the first cyclic product, lanosterol, will replace ergosterol, yeast's natural sterol. The conversion of squalene to lanosterol and the conversion of lanosterol to ergosterol are aerobic. Moreover, the anaerobic introduction of a methyl group at C-24, which occurs both in the presence as well as absence (Nes and Sekula, unpublished observations) of oxygen, is essential. Without the methyl group, yeast cells develop a pathology leading to death.

4. Anaerobic Biosynthesis in Aerobes

If we turn the problem around and examine the metabolism of obligate aerobes, the plot, rather than thickening, becomes more diffuse. Both anaerobic and aerobic types of cyclization of squalene occur in aerobes. This is shown in Figure 8. In the anaerobic route a proton attacks C-3 (steroid numbering), which bears one of the terminal double bonds of squalene. In the aerobic route the proton attack is delayed until after oxygen is introduced as an epoxy group on the same double bond (C-3 and C-4). The proton then attacks the oxygen. Otherwise, both routes are similar. The proton attack induces a positive charge on C-4, which then induces a flow of electrons toward C-4. This in turn, depending on the conformation adsorbed to the cyclase, induces closure of rings. These cyclizations are illustrated in Figure 8 for rings A and B., At the end of the cyclizations, which may or may not involve H or C migrations, the positive charge is left somewhere in the

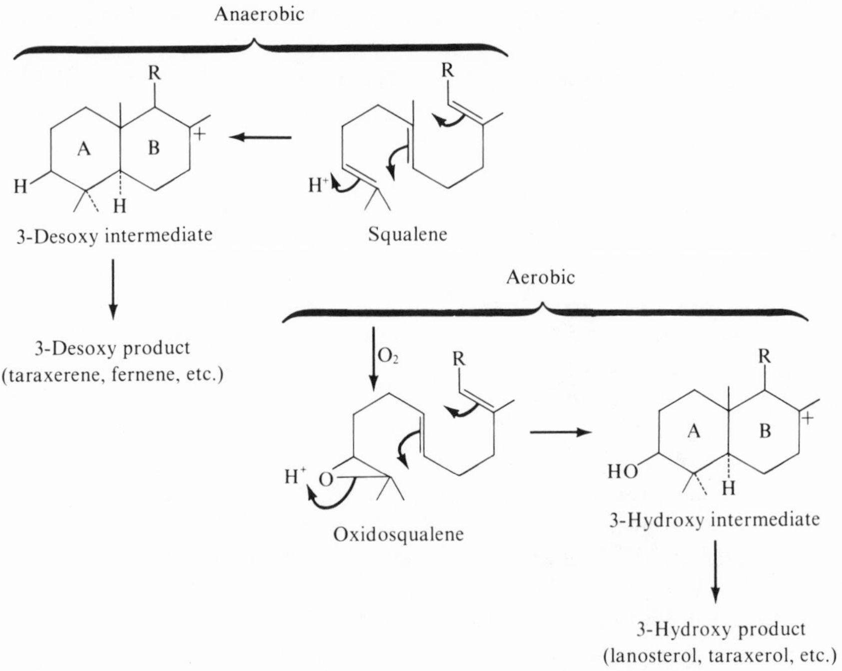

Figure 8. Anaerobic and aerobic cyclization of squalene. (The positive charge is removed, yielding the product by abstraction of OH⁻ from water or by loss of a proton with or without further condensations.)

molecule, as in the "intermediate" shown in Figure 8. Removal of the positive charge is accomplished either by abstraction of OH⁻ from water (producing an alcoholic hydroxyl group at that position) or loss of a proton (producing a double bond at that point). Examples of products derived from these processes are shown in Figure 9. Taraxerene and taraxerol illustrate the anaerobic and aerobic processes, respectively. Except for whether or not oxidosqualene is formed prior to protonation, what happens is exactly the same in the two cases. Diplotene and diplopterol represent the (anaerobic) routes in which the positive charge is eliminated by loss of a proton and abstraction of OH⁻, respectively. Diplotene and hopene I represent elimination of a proton with (hopene I) or without (diplotene) H migration, and in the case of fernene both H and C migrations have occurred. Zeorin and leucotyline represent the anaerobic and aerobic cyclization, respectively, with addition of OH⁻ and a subsequent aerobic hydroxylation at C-6. An especially interesting case is tetrahymanol, which is found in protozoa (Mallory *et al.*, 1963; Tsuda *et al.*, 1965). It is derived by the anaerobic route, but the folding pattern (conformation) of squalene at the moment of

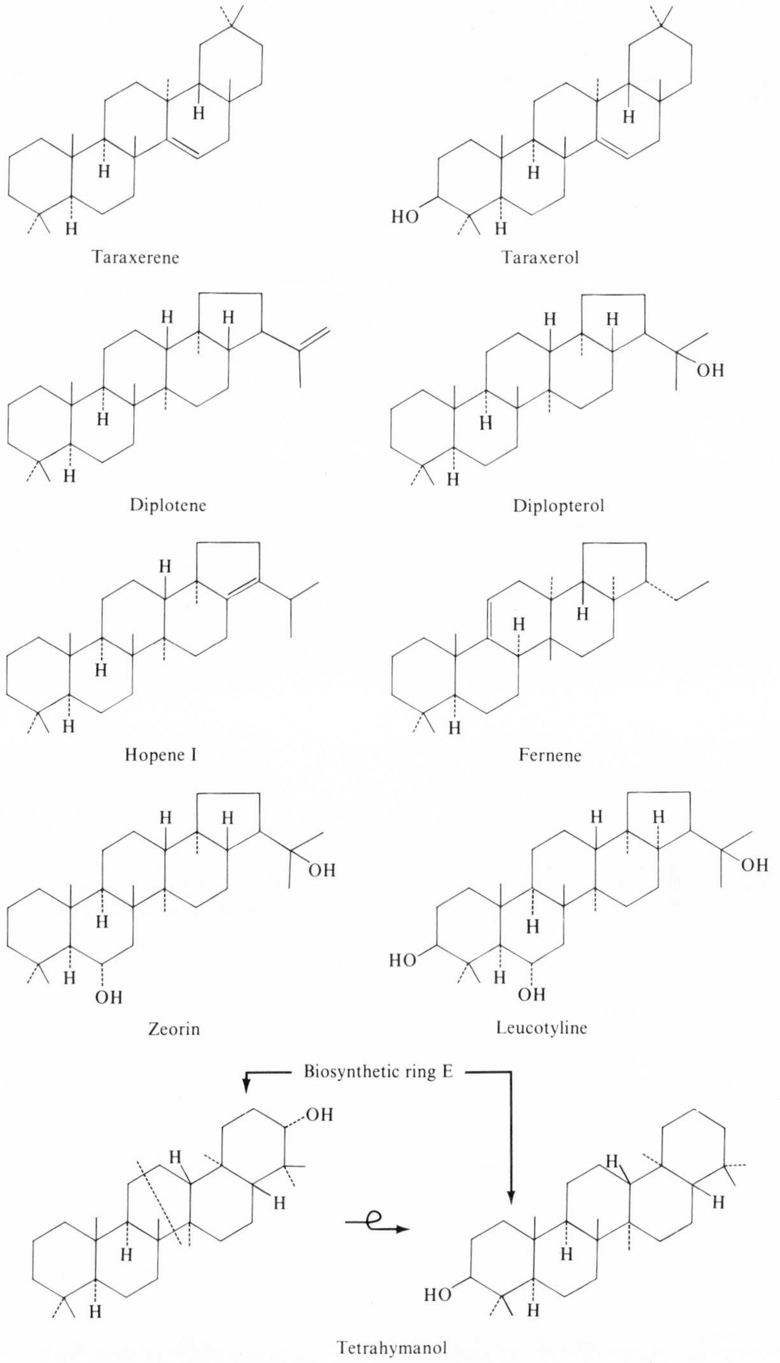

Taraxerene

Taraxerol

Diplotene

Diplopterol

Hopene I

Fernene

Zeorin

Leucotyline

Biosynthetic ring E

Tetrahymanol

Figure 9. Products of anaerobic vs. aerobic cyclization of squalene.

cyclization is such that, ignoring the OH group, one half of the molecule (on each side of the line in Figure 9) is identical to the other half except that one half is flipped over with respect to the other half. Consequently, if the whole molecule is rotated by 180° through the center, nothing is changed except which end is which. Since one end has an OH group, it will be seen from Figure 9 that the anaerobically introduced hydroxyl group on ring E assumes the position and stereochemistry of an aerobically introduced 3β-hydroxyl group on ring A. That tetrahymanol, fernene, and hopene arise anaerobically has been demonstrated by elegant experiments (Ghisalberti *et al.*, 1970; Zander *et al.*, 1970). The aerobic route has been the subject of even more extensive documentation (for a detailed review, see Nes and McKean, 1977). Lanosterol, cycloartenol, and β-amyrin are among many of the products derived aerobically.

From our present point of view, the interesting aspect of these cyclizations is that the anaerobic route occurs both in bacteria giving hopenes (de Rosa *et al.*, 1971a) and in very much higher forms of aerobic life. Taraxerene and taraxerol are formed both in the same (*Rhododendron falconeri*) and in different plants. The lichen, *Cladonia deformis*, for instance, has taraxerene, while taraxerol is found in 11 families of tracheophyte. Zeorin is present in *C. deformis* and in no less than 26 other species of lichen. Ferns, e.g., *Polypodium vulgare*, biosynthesize diplotene, hopene I, and fernene. Diplopterol is found in *Diplopterygium glaucum*. Nakai in the Gleicheniaceae. Tetrahymanol was first discovered in the protozoan *Tetrahymena pyriformis* and has since been identified in many other species of the genus. It also occurs in the plant kingdom. In addition to the references cited, the extensive review by Boiteau *et al.* (1964) should be consulted for a key to the literature on occurrence.

From the examples given it will be obvious not only that anaerobic biosynthesis occurs in aerobes but also that there is little association with evolutionary development. While it is true that there is the suggestion of a tendency toward the anaerobic route lower on the evolutionary scale, it remains to be seen from further work whether or not this is a real fact. As we shall examine in detail in a subsequent section, there is a better correlation between the type of molecule and anaerobic cyclization than there is between the appearance of organisms in the fossil record. One example here will suffice. The aerobic cyclization of squalene to a sterol (lanosterol or cycloartenol) has no anaerobic counterpart in any organism. To the best of our knowledge, neither 3-desoxylanosterol, nor 3-desoxycycloartenol, nor their hydrated analogs exist naturally, even though other pairs, e.g., taraxerol and 3-desoxytaraxerol (taraxerene), do exist. There is also a better correlation between anaerobic biosynthesis and other factors, e.g., the larger taxons, regardless of when they appeared on earth. In particular, no vertebrate is known which uses the anaerobic route. Vertebrates and vascular plants appear in the fossil record roughly at the same time, yet vascular plants do use the anaerobic route in some cases. Conversely, when we look at the aerobic route (to sterols) we find it occurring in the "primitive" prokaryotic blue-green algae and in man.

In the case of the protozoa there is an additional complication. *T. pyriformis*, which is aerobic, substitutes an anaerobic pathway to tetrahymanol for the aerobic route to sterol and also utilizes tetrahymanol in the way (membranously) that sterols are used (Holz and Conner, 1973). Furthermore, when sterol is present in the medium, *T. pyriformis* metabolizes it to the $\Delta^{5,7,22}$ trienol (Conner *et al.*, 1969; Nes *et al.*, 1971; Nes *et al.*, 1975a; Nes *et al.*, 1975b; Nes *et al.*, 1978b), tetrahymanol biosynthesis is depressed, and the sterol is used membranously.

Now, where did the receptors for sterol come from if the organism is so primitive as to use an anaerobic cyclization? There can be no doubt that the protozoan has proteins with active sites specifically complementary to sterols. Relatively small departures from the structure of cholesterol prevent its dehydrogenation to the $\Delta^{5,7,22}$ trienol. They include inversion of the configuration at C-20 and lengthening or shortening of the side chain (Nes *et al.*, 1978b; Nes, Joseph, Landrey, and Conner, unpublished observations). Changes in the structure of the side chain not only reduce or abolish dehydrogenation in the side chain but also do so in ring B, proving the existence of a very precise fit of the whole molecule into the active site. The dehydrogenases clearly have evolved with sterol "in mind." Similarly, the ability to inhibit tetrahymanol biosynthesis has structural correlates closely tied to sterol structure (Conner *et al.*, 1978). In addition to the problem of receptors, all protozoa do not biosynthesize tetrahymanol. An example is the sterol-requiring *Paramecium aurelia*, and of the various *Tetrahymena* species shown to biosynthesize tetrahymanol only *T. pyriformis* grows well in the absence of sterol (Holz and Conner, 1973). This mixed record suggests a complicated functional correlate and is difficult to explain on so simple a proposition as the development of oxygen in the distant past.

5. Sterol Biosynthesis and Phyla Evolution

Another way to examine the oxygen problem is to see whether phyla that the fossil record suggests developed early are more likely to lack an aerobic route than those that seem to have come later. The fossil record, at least for the older fossils, is best with true animals, and all of them, presumably including the ancient ones, are eukaryotic and aerobic. Do any use an anaerobic pathway to sterol-like materials? The answer is no, with the usual caveat "to the extent of our knowledge." Do any lack sterol biosynthesis with or without a sterol requirement? The answer here is yes, and all that lack the pathway have a requirement for sterol. In fact, the absence of sterol biosynthesis is very frequent in the animal kingdom.

A sterol requirement for insects was discovered many years ago by entomologists, and biochemists have recently broadened the matter by showing that not only insects but many other classes of arthropods have a block in the isopentenoid pathway (Nes and McKean, 1977). In the investigated cases, the block occurs prior to squalene, and none of the arthropods for which information is available carry the pathway to, say, the aerobic lanosterol or 14α-methyl stages and stops, thereby conferring a requirement for the finished product, a 4,4,14-trisdesmethyl sterol. Even though sterols are not biosynthesized, other isopentenoids, e.g., ubiquinone, are formed from MVA. Despite the absence of the sterol pathway, arthropods aerobically metabolize sterols to hormones, etc. (Bowers, 1978; Svoboda *et al.*, 1978).

If we look at the fossil record (Chapter 4, Section B) we find the arthropoda entered the time frame shortly after the base of the Cambrian, after which they flourished. They are thought to have been derived from the annelids. The annelids are of two sorts, marine and terrestrial. Neither has been extensively studied, but it is reported (Wooton and Wright, 1962; Walton and Pennock, 1972; Voogt *et al.*, 1975) that the marine polychaetes, *Nereis diversicolor*, *N. pelagica*, and *Arenicola marina*, convert MVA to sterols, while the earthworm (*Lumbricus terrestris*) carries the pathway only to squalene. This is a paradox. The earliest arthropods (trilobites), presumably lacking aerobic formation of sterols, were marine animals. Yet, extant marine annelids, (e.g., *N. diversicolor*), presumably representing the line to arthropods, seem to use an aerobic pathway, while the terrestrial annelid (*L. terrestris*), which should have become better adapted to aerobiosis than the one in the oceans, has a block. The situation is made worse by the phylum Mollusca.

In the fossil record molluscs predate the arthropods. If anything, anaerobic modes of metabolism should have been more ingrained in them than in the arthropods, but the reverse is true. Biosynthesis of sterols is a common property of molluscs (Voogt, 1975; Teshima and Kanazawa, 1974; and references cited). On the other hand, the phylum Coelenterata is beleved to be still more ancient than the molluscs or arthropods, and it should be a still better marker of anaerobiosis. Indeed, a kind of metabolism of mevalonic acid known in some bacteria has been observed, i.e., retrocondensation to acetate. Label from $[2\text{-}^{14}C]$-MVA proceeds strongly in various coelenterates, e.g., the sea anemone (*Anemonia sulcata*) (Voogt *et al.*, 1974), to phospholipids, but little if any into sterols (for a further discussion, see Nes and McKean, 1977). The platyhelminthes (flatworms and flukes) also fail to biosynthesize sterols (Frayha, 1974; Smith *et al.*, 1970), and these organisms are supposed to predate arthropods and molluscs. The flatworms and flukes are acoelomate (no body cavity). Higher animals are eucoelomate (possess a developed body cavity or coelom). Intermediate are the unsegmented round worms in the phylum Nemathelminthes. The round worms carry the pathway to lanosterol but not further, i.e., aerobic but sequentially incomplete aerobic metabolism occurs, leading to a requirement for an exogenous source of sterol (Rothstein, 1967; Willett and Downey, 1974). The sponges (phylum Porifera), originating in the Cambrian or earlier, are less satisfactory at the moment, since some do and some do not biosynthesize sterols (Walton and Pennock, 1972; de Rosa *et al.*, 1973).

6. Sterol Biosynthesis in Prokaryotes

Prokaryotes are so named owing to their lack of membrane-bound chromosomal material. Typically they also lack mitochondria and other organelles that are usually associated with membranes. However, the absence of so-called "double membrane" organelles (Weier *et al.*, 1965, 1966) is a quantitative rather than an absolute phenomenon. For instance, in blue-green algae and photosynthetic bacteria the photosynthetic lamellae (thylakoids) have recently been shown through ultrastructural and other methods to be organized aggregates. Although lacking their own double membrane that would constitute a "true" chloroplast, the photosynthetic apparatus is certainly particulate as seen by electron microscopy (Bisalputra, 1974; Arntzen and Briantais, 1975). Furthermore, in bacteria, e.g., *E. coli,* some sort of specialized architecture on the inside of the cytoplasmic envelope binds the apparatus for functions such as protein synthesis and electron transport, which in eukaryotes is microsomal (Gelman, 1975). Prokaryotes are thus not simply a bag with biochemistry occurring randomly in the solution inside. Even these least organized of cells are still highly organized at a particulate level. They contain at least one multifunctional membrane (limiting envelope) in the nonphotosynthetic case and more than one membranous system in the photosynthetic case, i.e., in the thylakoids. As pointed out earlier, it has been suggested that thylakoids do not exist in prokaryotes as double-membrane entities. The latter view has been one of many determining factors for systematics and phylogeny of prokaryote evolution. Recently, however, investigators in the membrane field have provided convincing evidence that biological membranes do not exist as alternating layers of proteins and phospholipids; the membrane is a quasicrystalline-fluid matrix and asymmetric (Jain and White, 1977). Consequently, the strict interpretation of electron micrographs of cell structure at a molecular level and their relationship to evolution is suspect.

The importance of the above discussion of particulate organization is that the type of subcellular organization or lack of it has led to the supposition that prokaryotes have no sterols. The requirement for oxygen in steroid biosynthesis coupled with the existence of the anaerobic life of some bacteria seemed to correlate with this supposition. The idea was given added weight by very early reports that bacteria do not contain sterols and a more recent one (Levin and Bloch, 1964) that blue-green algae are sterol-less. The real facts are somewhat different.

Certain bacteria use "swamp gas" (methane) as a source of carbon. They frequently live in conjunction with bacterial methanogens in shallow waters. *Methylococcus capsulatus* unquestionably biosynthesizes sterols. This was first discovered by Bird *et al.* (1971) and later reinvestigated and refined in another laboratory (Bouvier *et al.*, 1976). The sterols, which contain a $\Delta^{8(14)}$ bond, are not the usual sort. Similarly, Schubert *et al.*, (1968) found *Azotobacter chroococcum* contained Δ^7 sterols. In both bacteria there was a high concentration of intermediates retaining one to three of the methyl groups at C-4 and C-14, and farnesol and squalene were found in *M. capsulatus*. The unusual sterols, the presence of intermediates, and in the case of *M. capsulatus* the repetition of the work in a second laboratory (and one versed in sterol isolations) can leave little doubt that these two prokaryotes possess the capacity for aerobic biosynthesis. Another methylotroph, *Methylobacterium organophilum*, also biosynthesizes squalene and three sterols one of which is believed to be 22-dehydrolanosterol (Patt and Hanson, 1978). Sterols have also been reported from *E. coli* (Schubert *et al.*, 1964), *Streptomyces olivaceus* (Schubert *et al.*, 1967), and *Micromonospora* sp. (Fiertel and Klein, 1959), but in the case of the first two the results are questionable, since cholesterol and other common Δ^5 sterols were found. Δ^5 Sterols are very difficult to exclude from the environment. We have recently reinvestigated *E. coli* (Nes and Nes, unpublished observations). No sterols could be detected, although peaks in the mass spectrum corresponding to cholesterol and its 24-methyl and 24-ethyl derivatives were found just at or slightly above the level of the spectral background. We also have been unable to detect sterols in two genera of photosynthetic bacteria (*Chromatium vinosum* and *Rhodopseudomonas spheroides*) (Nes and Frasinel, unpublished observations), and others have recently failed to find sterols in several genera of nonphotosynthetic bacteria.

The ability of blue-green algae to biosynthesize sterols is more consistent (Nes and McKean, 1977). All eight genera, which have been examined in several laboratories, contain sterols. In at least one case (*Phormidium luridum*) squalene was observed, and in most cases the sterols present were not the usual ones (the homologous cholesterol series), contraindicating contamination. The blue-greens clearly possess the aerobic pathway of steroid biosynthesis.

The record from other prokaryotes is mixed. Some bacteria reportedly carry the pathway only to squalene, but others are said not even to contain this hydrocarbon. If things were simple, which clearly they are not, we would have liked to have seen bacteria always carrying the pathway to squalene and stopping—or perhaps doing as *Halobacterium cutirubum* does, reduce squalene. In the case of the bacteria-like prokaryotes lacking a cell wall (mycoplasmas), none is known to biosynthesize either squalene or sterols. However, some of the mycoplasmas ("choleplasmas" recently classified in the genus *Mycoplasma*) have an absolute requirement for sterol, which becomes a constituent of the cytoplasmic membrane. An example is *Mycoplasma gallinarum.*

The inability of choleplasmas to biosynthesize sterol while having a sterol requirement has many counterparts in eukaryotic organisms from protozoa to true animals (Sections B-4 and B-5) and is obviously characteristic of neither prokaryotes nor eukaryotes. The most likely reason why cells do or do not biosynthesize or do or do not require sterols seems to us to be more of an ecological–functional problem than one having to do with the origin of oxygen. If the cells must have sterol or something like it (carotenol, tetrahymanol, etc.) for their function, they have no choice but to make an appropriate molecule or acquire it from another cell. The choleplasmas, certain of the protozoa, insects, etc., take the latter choice or die. The eukaryotic *Tetrahymena pyriformis* (Section B-4) and another group of mycoplasmas, the genus *Acholeplasma*, such as *A. laidlawii*, make the former choice. Acholeplasmas biosynthesize an alcoholic carotenoid (aerobically?), which the choleplasmas do not. The carotenol, which in a certain conformation could conceivably mimic the polycyclic structure of sterols (Nes, 1974), is thought to take the place of the membranous sterol, explaining why the two very similar organisms have such different nutritional requirements.

Before leaving prokaryotes we should like to point out that some bacteria metabolize steroids. Examples are the reduction of cholesterol by *Eubacterium* sp. and other fecal bacteria; the isomerization of Δ^5-3-ketones to Δ^4-3 ketones by a steroid-induced enzyme in *Pseudomonas testosteroni;* desaturation of 3-keto steroids at C-1(2) or C-4(5) by *Corynebacterium, Pseudomonas,* and *Streptomyces;* and aromatization of ring A with or without cleavage at C-9(10) by *Pseudomonas, Arthrobacter,* and *Mycobacterium*. These bacteria at some point in their evolution "learned" to use "advanced" steroids. In reverse, some bacteria seem to have "forgotten" how to operate the very earliest part of the lipid pathway. For instance, the anaerobe *Lactobacillus acidophilus* requires acetate itself. In an appropriate medium (which presumably contains fatty acids) mevalonic acid will replace the acetate, leading to biosynthesis of the isopentenoid "carbohydrate carrier lipid" (CCL), which is required in cell-wall formation. The bacterium thus lacks steps such as glycolytic ones, which yield acetate, but has enzymes for the rest of the (anaerobic) pathway to CCL. These "gains" and "losses" of enzyme, which can be documented in many another case, pose very difficult problems to us. If we say the "gain" of a vertebrate type of enzyme (Δ^5 isomerase involved in adrenocortical and gonadal-hormone biosynthesis) by *Pseudomonas* is an adaptation, why is the enzyme in *E. coli* for β,γ-desaturation in fatty acid biosynthesis a vestige of Precambrian times and not just also an adaptation?

7. The Possibility of Recapitulation

The problem of recapitulation in a highly differentiated eukaryote has been investigated in the authors' laboratory with seeds of tracheophytes (McKean and Nes, 1977; Nes *et al.*, 1967). It is not unreasonable to suppose that early in embryonic development carbon might traverse the isopentenoid pathway only through its anaerobic portion, leading to an accumulation of squalene. Sterol that might be needed by the embryo could be deposited by the mother, who "knows" the embryo is actually not fully primitive. Cholesterol, for instance, in a hen's egg actually comes largely from the hen. In the case of plant seeds, the angiosperm *Pisum sativum* (peas) was shown to accumulate squalene strongly during the first 24 hr of germination, as the hypothesis predicts. No sterol at all was formed from squalene and only a small amount of pentacyclic products (α- and β-amyrin). The seed was shown to have a great deal of both sterol and pentacycle, in further agreement with the postulate. As germination proceeded the metabolic systems changed such that during the second to fourth days squalene was aerobically cyclized to the amyrins, and with the appearance of leaves on about the fifth day not only aerobic cyclization to cycloartenol but further aerobic demethylation, etc., occurred to convert cycloartenol to Δ^5 sterols (24-methyl- and 24-ethylcholesterol). The steady-state concentration of squalene fell during this period to near zero.

We have also demonstrated accumulation of squalene in seeds of the gymnosperm *Pinus pinea* (the stone pine) (McKean and Nes, 1977). As in the case of peas, the stone pine seeds were allowed to absorb labelled MVA in the water they require to begin the process of germination. However, this labelled different parts of the two kinds of seed. Peas are essentially all embryo, and the embryo, especially the cotyledon, which constitutes most of the weight and surface, was the site of labelling and metabolism. However, in the pine the embryo is encapsulated by maternal tissue (the endosperm) which represents half or more of the weight of the seed. The result is primarily labelling of and metabolism in the endosperm. The endosperm is not connected to the embryo, and the two tissues can be separated readily and each examined. To get the label directly into the embryo, we germinated the seed in water and then directly fed the embryo an MVA solution through the roots. As anticipated from the work with the peas, the pine embryos (possessing much sterol) accumulated squalene early in their development and only later developed marked capacity to convert the hydrocarbon to sterols. As sterol formation occurred, no other intermediates (cycloartenol, etc.) accumulated in place of squalene, suggesting a single block at the first aerobic step (epoxidation).

Despite the consistency of this result in two quite different plants, we found exactly the same phenomenon (early accumulation of squalene with later development of sterol formation) in the pine endosperm and were able to show that no transfer from one tissue to the other occurred. The accumulation of squalene in the tissue (endosperm) that has (haploid) maternal genetics seems to weigh heavily against an interpretation based on recapitulation, since one would prefer to invoke this idea with a zygote. Fertilization, of course, establishes time zero. Probably the explanation for the accumulation lies in the hydrophobic character of squalene. One can imagine it regulating the distribution of the water taken in during the early phases of germination. Unusual concentrations of squalene are also found in mature elasmobranch livers, which can hardly be supposed to represent recapitulation. In fact, evidence exists linking squalene and other lipids in fish to a regulation of their buoyancy.

8. Additions of Aerobic Steps

We have previously referred to the fact that the lipid pathways are extended by additions of O_2-dependent reactions in aerobic organisms. The purpose of this subsection is to summarize what these are and to see whether there is any association with an evolutionary hierarchy. Our own view is that in some cases there is such an association but that its relationship to the supposed development of our atmosphere seems tenuous. One of the best examples is in the vertebrate bile acid pathway, but its development occurred long after adequate oxygen was available for the Cambrian "explosion" of life. A more likely explanation lies in the evolution of complexity per se, which simply took time to achieve, especially when it occurred in concert with the development of appropriate functions for the products of the aerobic reactions. Figure 10 summarizes the aerobic additions. It should be compared with Figures 5 and 6, which summarize the basic anaerobic routes. The details of these O_2-dependent reactions are beyond the scope of this chapter. The interested reader should consult the extensive reviews that already exist, e.g., Nes and McKean (1977) and Hamberg *et al.* (1974).

The first and principal addition in the sterol and related pathways is the aerobic epoxidation of squalene to give oxidosqualene, which is then cyclized

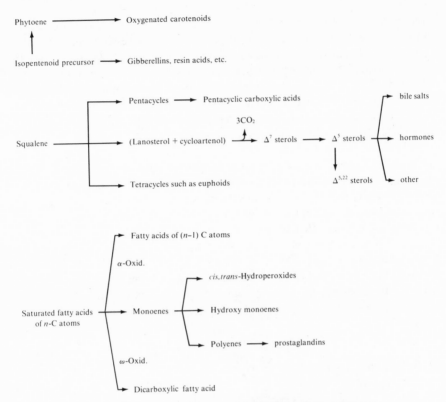

Figure 10. Examples of O_2-dependent additions to lipid pathways.

to lanosterol and other tetracycles (cycloartenol, euphol, etc.) as well as pentacycles (the amyrins, etc.). In the sterol pathway, but not in the pentacyclic and euphoid pathways, the product of cyclization (lanosterol and cycloartenol) proceeds through O_2-dependent demethylations at C-4 and C-14 to the functional sterols, e.g., cholesterol and sitosterol. The number and arrangement of double bonds is also partly O_2 dependent. The Δ^7 bond is derived anaerobically from a Δ^8 bond. The Δ^{24} bond and double bonds derived from its alkylation, e.g., $\Delta^{24(28)}$ and $\Delta^{25(27)}$, are either retained as such or metabolized by reduction, but the introduction of a Δ^5 or Δ^{22} bond is aerobic. In those bacteria that clearly make sterols, the Δ^5 bond is not introduced. The pathway stops at the Δ^7 or $\Delta^{8(14)}$ stages. When we look at eukaryotes, we see a strong tendency toward completion of the pathway with introduction of the Δ^5 bond. Both animals and higher plants biosynthesize cholesterol or its 24-alkyl derivatives. Many exceptions exist, though. For instance, the highly evolved plant family Cucurbitaceae (pumpkins, cucumbers, etc.) makes only Δ^7 sterols (Nes *et al.*, 1977a), and mammalian skin is unusually rich for a vertebrate tissue in the Δ^7 analog (lathosterol) of cholesterol. Among algae, Δ^5 sterols are also more common, but there are five species in the genus *Chlorella* that are exclusively Δ^7 producers, and one each in *Scenedesmus*, *Oocystis*, and *Hydrodictyon*. Among the problems with using data of this sort as a marker of the development of an oxygen atmosphere is that the absence of an aerobic step in one organism can be paralleled by the absence of an anaerobic step in another. Thus, in the genus *Chlorella*, which includes the Δ^7 species, there are also species that introduce the Δ^5 bond but fail to reduce the Δ^7 bond. This leads to the presence of a $\Delta^{5,7}$ sterol. Why should we choose the lack of one O_2-dependent reaction (Δ^5 introduction) and ignore the lack of a reduction (saturation of the Δ^7 bond)?

Just as there is no necessary correlation of presence or absence of these

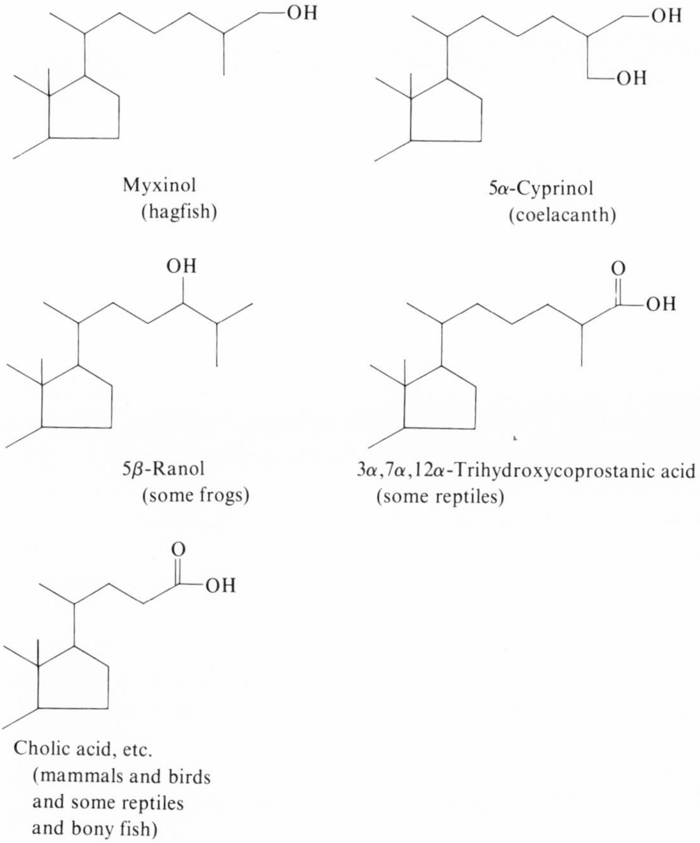

Myxinol
(hagfish)

5α-Cyprinol
(coelacanth)

5β-Ranol
(some frogs)

3α,7α,12α-Trihydroxycoprostanic acid
(some reptiles)

Cholic acid, etc.
(mammals and birds
and some reptiles
and bony fish)

Figure 11. Side chains of steroids in bile salts.

desaturations in a single genus such as *Chlorella*, there is none necessarily in a given species. The introduction of a Δ^{22} bond serves as an example. In anaerobic yeast supplied, for instance, with 7-dehydrocholesterol or 24β-methylcholesterol, no Δ^{22} bond is introduced (Nes and Sekula, unpublished observation), while it is formed under aerobic conditions, as shown by yeast's natural sterol (7,22-bisdehydro-24β-methylcholesterol, ergosterol) and various incubations with 22-dihydrosterols that yield ergosterol. We see in yeast a case where two O_2-dependent dehydrogenations (for the Δ^5 and Δ^{22} bonds) occur under aerobic conditions, but one reduction ($\Delta^{5,7}$ to Δ^5) does not occur. Conversely, animals commonly reduce the $\Delta^{5,7}$ diene but do not introduce the Δ^{22} bond, and in photosynthetic systems there is a complicated pattern of Δ^{22} introduction depending on species, tissue, etc., which bears no relationship to an evolutionary hierarchy.

Perhaps a better relationship between the evolution of a taxon and the addition of O_2-dependent reactions is to be seen in the bile salts of vertebrates. There can be little doubt from the fossil record that fish and reptiles came earlier than mammals. We also now know that the conversion of cholesterol to the bile acids used for bile salt formation in mammals is aerobic. The steps involve hydroxylation of one of the terminal methyl groups in the side chain, oxidation of the primary alcohol to the carboxyl stage, hydroxylation at C-24, and an oxidative retroenolate condensation with cleavage of the bond between C-24 and C-25 releasing the bile acid and propionic acid. While the intermediates do not accumulate normally in mammals, in lower vertebrates they assume the place of the mammalian acid. In particular, the hagfish, which is in the most primitive group of vertebrates (Cyclostomata), has a 26- or 27-alcohol as does a member of the Chondrichthyes. In the more advanced bony fish one then finds the usual acids, e.g., taurocholic acid, although the coelacanth has a 26,27-diol. In land animals a similar progression exists. While many reptiles have cholic acid (a C_{24} acid), a C_{27} acid is found in some species (Figure 11).

Another O_2-dependent steroid pathway frequently found in the higher reaches of the evolutionary hierarchy is the sequence, or more properly the sequences, leading to steroid hormones. In fungi, insects, vertebrates, and plants, hormones of many sorts are formed from sterols by various hydroxylations on saturated carbon atoms that clearly require oxygen. Not only are hydroxylations performed, but as in the bile acid pathway, carbon atoms in the side chain are removed in many well-studied cases. Progesterone, cortisol, and aldosterone, for instance, result from cleavage at C-20(22), and the estrogens, androgens, and dehydroepiandrosterone are derived by cleavage at C-17(20). In fungi, carbon atoms are not removed; rather, they are added (alkylation at C-24), forming such hydroxylated hormones as antheridiol. Steroidal hydroxylations go beyond just the formation of hormones.

There exist a great variety of steroids in both plants (sapogenins, etc.) and animals (bufenolides, etc.) that are derived by hydroxylations with or without subsequent reactions. Essentially every carbon atom in the sterol molecule has been shown to undergo hydroxylation in one or another organism, and some oil microorganisms use sterols as their sole source of carbon.

Similarly, in the fatty acid pathway there are examples of aerobic additions to the basically anaerobic pathway. Aerobic desaturation of fatty acids is carried out in *Bacillus megatorium*, *Corynebacterium diphtheriae*, *Micrococcus lysodeikticus*, and *Mycobacterium phlei* (Hitchcock and Nichols, 1971, and references cited). Bacteria usually do not contain 1,4-polyunsaturated fatty acids, although in principle they could multiply desaturate even anaerobically. *E. coli*, for instance, could carry out β,γ desaturation more than once. This would lead in sequence to a conjugated set of double bonds. To prevent conjugation, several alternatives are conceivable. An all-*cis*-$\Delta^{9,12}$ diene, for instance, would result from a *cis*-β,γ-dehydration followed by a *cis*- (instead of *trans*-) α,β-dehydration after the next C_2 addition. The reason why this is not done is unclear, but *Bacillus licheniformis* introduces two double bonds into palmitate by O_2-dependent reactions to give a diene of the 1,6 type, 5,10-hexadecadienoate (Fulco, 1970). Fatty acids with the 1,5 type of dienic structure are found in *Mycobacteria* (Asselineau *et al.*, 1972), but the mechanism is unknown (Fulco, 1977). In contrast to the apparently rare occurrence of polyunsaturation in bacteria, the blue-green algae commonly desaturate aerobically to give $C_{18:3}$ and $C_{18:4}$ (Wood, 1974). The eukaryotes also frequently add an O_2-dependent desaturation leading not only to the monoenes but also to the polyenic fatty acids, e.g., $\Delta^{6,9}$, $\Delta^{9,12}$, and $\Delta^{6,9,12}$. Additional O_2-dependent metabolism of fatty acids is also known. α-Oxidation (actually, α-hydroperoxidation) occurs in higher plants (and in mammalian brain and liver) with the formation of several products: the acid, aldehyde, or fatty alcohol with one less carbon atom and the α-hydroxyacid with the original number of C atoms (Stumpf, 1976). O_2-Dependent hydroxylation of oleyl CoA (Δ^9) at C-12 to give ricinoleyl CoA occurs in the castor bean, but as in so many other cases alternatives occur. Evidence exists for the formation of ricinoleic acid in mycelial cultures of the fungus *Claviceps purpurea* by a reaction not involving O_2 that is believed to involve the addition of HOH or HOCOR (followed by hydrolysis in the latter case) to the Δ^{12} bond of linoleic acid. This mechanism is obviously representative of an anaerobic route.

ω-Oxidation is another example of O_2-dependent metabolism of fatty acids that occurs widely in eukaryotes from man and other animals to yeast but again also occurs in bacteria, such as *Pseudomonas oleovorans*. In higher plants, especially well studied in soybeans but occurring widely, *cis*-unsaturated fatty acids proceed to *cis,trans*-dienic hydroperoxides in the presence of O_2 and a "lipoxygenase," and the fungal occurrence of ergosterol 5,8-peroxide, etc., probably has a related origin. In animal tissue, studied in sheep vesicular gland, guinea pig lung, rabbit kidney, and many other tissues, peroxides have been implicated in cyclization of fatty acids (Samuelsson *et al.*, 1975; Lands *et al.*, 1977). Arachidonic acid is converted in this way into prostaglandin E_2 (PGE$_2$). Analogously the $\Delta^{8,11,14}$-$C_{20:3}$ acid yields PGE$_1$, and the $\Delta^{5,8,11,14,17}$-$C_{20:5}$ acid gives PGE$_3$. The cyclization is an O_2-dependent reaction resembling what happens in the presence of lipoxygenase. A double bond is attacked by O_2, and the resulting peroxy group as a free radical then attacks another double bond, inducing it to react with the first double bond, closing a ring to give PGG$_2$ (the "mother prostaglandin"), which now bears a peroxide bridge. The latter is rearranged to give the hydroxyketocyclopentane moiety of the PGE series. An additional hydroxylation (actually occurring early in the sequence) completes the structures (Pace-Asciak, 1977).

Another way of viewing aerobic additions to the basic anaerobic route is with respect to the extent of desaturation of fatty acids. As one proceeds up the evolutionary scale in time and presumably in complexity we might expect to find an increasing number of double bonds and increasing chain length. Actually, such is not the case. Blue-green algae have as many as 18 C atoms with four double bonds (Wood, 1974). Eukaryotic algae contain as much as 22 C atoms in a fatty-acid chain with six double bonds (Erwin, 1973). However, in mosses we reverse ourselves and find C_{20} fatty acids with as many as four double bonds (Gellerman, *et al.*, 1972), and still higher at the tracheophyte level the chains do not exceed C_{18} with three double bonds (Galliard, 1973). In a sense, at the top of the scale is man. While requiring C_{18} and C_{20} polyunsaturates, he only carries out *de novo* biosynthesis of $C_{18:1}$.

C. SUMMARY

Despite our initial enthusiasm in the discovery that lipid biosynthesis is essentially anaerobic even to the point of yielding unsaturated fatty acids for membranes in some prokaryotes with aerobic desaturation occurring in eukaryotes, the flaws in the argument for a primitive anoxygenic environment point less toward than away from authenticity. Insofar as the bacterial anaerobic formation of unsaturated fatty acids is concerned, some bacteria do not even make fatty acids for their membranes (Chapter 6). Recapitulation of anaerobiosis cannot be adequately documented either in single-celled or differentiated organisms. The anaerobic formation of unsaturated fatty acids is not shifted to an aerobic synthesis by the presence of O_2 in bacteria, nor is aerobic desaturation shifted to the anaerobic pathway in yeast by deletion of

O_2. In the sterol pathway, recapitulation is suggested by the embryonic accumulation of squalene in seeds, but similar accumulation occurs in maternal tissue. Moreover, the absence of oxygen fails to induce an anaerobic pathway of polycyclic isopentenoid biosynthesis in yeast, and a typical anaerobic product (tetrahymanol) fails to replace sterol in this organism. When one examines primitive vs. advanced subcellular development, a mixed not consistent result is found. While some bacteria do not carry the isopentenoid pathway aerobically to sterols, some do. Similarly, some eukaryotes do and some do not biosynthesize sterols. Except for a few cases where a sterol-like material is formed endogenously, all eukaryotes which fail to biosynthesize sterols require an exogenous source of sterol. However, this phenomenon is independent of subcellular organization, since some prokaryotes (*Mycoplasma* sp.) also require but do not biosynthesize sterols. At a higher level of organization, the development of the coelom does correlate with introduction of an aerobic pathway but only if the Arthropods and Annelida are neglected. Acoelomates (Platyhelminthes) possess the entire sterol pathway, animals with a pseudocoelom (Nemathelminthes) carry the pathway aerobically only to lanosterol, and eucoelomates e.g., vertebrates, have the entire pathway. Turning to anaerobic lipid biosynthesis in eukaryotic aerobes, we find that aerobes are not devoid of anaerobic modes of cyclization of squalene. Whether they do or do not possess an anaerobic route is independent of whether they are single-celled protozoa or highly differentiated vascular plants, suggesting that the level of development is not determining. When the fossil record is considered, no better correlation is observed. Early lines, e.g., the molluscs, biosynthesize sterols, while later lines, e.g., the arthropods, do not. Marine annelids do while terrestrial annelids do not biosynthesize sterols.

To use the available data as an argument for a primitive anoxygenic environment seems to us to be a risky business, especially in view of the highly questionable nature of the geochemical and astrophysical evidence. Our own view, while tentative, is that the principal determinant of the use of aerobic vs. anaerobic pathways more likely resides in the adaptation of the cell in question to its particular ecological niche. This is not to say that we think the lipid data preclude a lack or paucity of oxygen early in earth's development. We simply feel the available data do not strongly support it. On the other hand, the data do suggest that most subtaxons in a given large taxon utilize aerobic or anaerobic pathways to a similar degree. This is most well documented in the sterol pathway, allowing the generalization, for instance, that vertebrates do and arthropods do not have an aerobic pathway to sterols. But in other phyla, e.g., porifera, even this sort of interpretation suffers from inconsistency.

CHAPTER 6

The Temperature and pH Problem

A. MAXIMUM LIMITS FOR ORGANISMS

Since there is reason to believe that the core of the earth is currently hot and that genesis of the earth itself from "dust" might have generated heat, the primitive surface of our planet is sometimes also thought to have been hot. Whether or not it actually was is another matter, about which geologists do not agree today. The origin of both our planet and the other planets is much more speculative than frequently supposed, and the complexity of the problem has been dramatically emphasized by very recent studies of Mars and Venus. However, on the supposition that the surface was hot, some investigators have regarded thermophilic bacteria as a likely kind of organism to first have arisen on the earth.

As a result of an extensive study carried out by Brock primarily in Yellowstone National Park in the decade from 1965 to 1975, a great deal of information on the ability of life forms to exist at high temperatures has now become available. His work, summarized well in a recent book (Brock, 1978), stimulated others to examine the lipids in some of these organisms which we shall review in the next subsection, but equally as interesting he was able to set some limits on the maximum temperatures and pH's tolerated by various organisms.

From Table 14 it will be seen that prokaryotes (blue-green algae, photosynthetic bacteria, and anaerobic and aerobic nonphotosynthetic bacteria) will tolerate temperatures well in excess of 60°, while this is the upper limit for eukaryotic microorganisms (Tansey and Brock, 1972; Brock, 1978). With the development of substantial cellular differentiation the maximum

Table 14. Maximum Temperatures for Various Life Forms[a]

Organism	Temperature
Prokaryotes	
Blue-green algae	45° –74°
Chroococcales	
Synechococcus	74°
Chamaesiphonales	52° –54°
Oscillatoriales	45° –60°
Oscillatoria	53° –60°
Phormidium	46° –60°
Nostocales	52° –64°
Calothrix	52° –54°
Photosynthetic bacteria	57° –73°
Chlorobiaceae	
Chloroflexus	70° –73°
Chromatiaceae	
Chromatium	57° –60°
Nonphotosynthetic bacteria	50° –90°
Spore-formers	
Bacillus	50° –82°
Clostridium	65° –75°
Lactic acid bacteria	53° –65°
Actinomycetes	55° –75°
Methanogens	75°
Methane-oxidizing bacteria	
Methylococcus capsulatus	55°
Sulfur-oxidizing bacteria	55° –90°
Sulfolobus acidocaldarius	85° –90°
Sulfate-reducing bacteria	85°
Mycoplasma	
Thermoplasma acidophilum	65°
Gram-negative aerobes	79° –85°
Eukaryotes	
Animals	
Fish	38°
Insects	45° –50°
Crustaceans	49° –50°
Vascular plants	45°
Mosses	50°
Protozoa	56°
Algae	55° –60°
Fungal microorganisms	60° –62°

[a]After Brock (1978).

temperature decreases further to 38–50°. The highest temperature photosynthetic organisms can tolerate is 74° (*Chloroflexus* and *Synechococcus*), while nonphotosynthetic organisms can sustain temperatures somewhat higher. The sulfate-reducing bacteria and gram-negative aerobes can reach 85°, and the sulfur-oxidizing bacteria and mycoplasmas will allow as much as 90°. The maximum temperatures do not correlate strongly with anaerobic vs. aerobic modes of life, and the absence of photosynthesis is more conducive to a tolerance of high temperatures than its presence. If the surface of the earth were at first hot (90+°), anoxygenic, and aqueous, the data do permit one to say that nonphotosynthetic, anaerobic bacteria could conceivably have initially existed with the subsequent rise of photosynthetic prokaryotes when the temperature fell 20 or so degrees. There is in this, though, a certain amount of choosing of data to fit the model. The information on thermophilic organisms also permits aerobic bacteria to have evolved to 85° prior to the evolution of O_2-producing blue-green algae at 74°! Moreover, blue-green algae also grow at very cold temperatures. They are the dominant plant life in the frigid lakes of the Antarctic (Fogg *et al.,* 1973), so their presence on the primitive earth tells us little about the temperature. More work and thought is obviously called for.

In the course of his studies Brock (1978) also provided us with interesting information on the effect of pH. In particular, the lower pH limit for organisms shows a complicated pattern. This is illustrated by the following examples (pH, organism): 4, fish and blue-green algae; 3, vascular plants, mosses, bacteria (*Bacillus*) and fungi (*Streptomyces*); 2, insects and protozoa; 1–2, some eukaryotic algae, e.g., *Euglena, Chlamydomonas,* and *Chlorella;* 0.8, some bacteria, e.g., *Sulfolobus;* and O, some eukaryotic algae, e.g., *Cyanidium,* and fungi, e.g., *Acontium.* An acid environment is clearly tolerated by a diverse group of living things. Except for the gross fact that vertebrates are more sensitive to the lower pHs than are some prokaryotes, the ability of an organism to withstand strong acid is certainly not simply determined by where it stands on the evolutionary hierarchy. Some eukaryotic algae and fungi will tolerate as strong an acid environment as will bacterial acidophiles, while other bacteria can live at no more acid a pH than do vascular plants. Such information correlates better with ecological adaptation than with a relationship to prior conditions. Perhaps more interestingly in our quest for origins, prokaryotic blue-green algae do not exist in waters which exceed a pH of 4, while their eukaryotic relatives (*Rhodophyta, Bacillariophyta* (diatoms), *Euglenophyta,* and *Chlorophyta*) do. This suggests that more or less neutral waters may have predated acidic ones and not the other way around, or at least that the first seas containing O_2-evolving organisms were not very acid. In turn perhaps we can exclude thermophilic acidophiles from the possible roster of those nonphotosynthetic bacteria that could have come earlier than the blue-greens. In this connection, the formation of stromatolitic mats occurring through the combined action of photosynthetic bacteria and blue-green algae in Yellowstone takes place in neutral to alkaline waters, and the existence of Precambrian stromatolites composed of limestone indicates the ocean, or at least that part of it which produced the stromatolites, was not significantly acid (otherwise limestone could not exist).

B. LIPIDS OF THERMOPHILES

1. *Cyanidium caldarium*

This unicellular, eukaryotic alga with mitochondria, a single chloroplast, and a membrane-bounded nucleus survives at 55–60° and is found in hot, acidic (pH > 4) springs. At temperatures above 40°, it is the only photosynthetic component. Disagreement exists as to how to classify it. In particular, various morphological and biochemical aspects point toward blue-green algae, while others suggest several kinds of higher algae, especially the Rhodophyta. It contains chlorophyll *a* as its sole chlorophyll, c-phycocyanin, allophycocyanin, β-carotene, lutein, zeaxanthin, and no myxoxanthophyll.

Ikan and Seckbach (1972) showed the presence of sterols by gas–liquid chromatography, mass spectrometry, and ultraviolet spectroscopy. Although sufficient information was not reported to allow determination of several important parameters, notably configurations, the data are consistent with the presence of the homologous cholesterol series (24-H, 24-CH$_3$, and 24-C$_2$H$_5$ in increasing order of amounts), the related 7-dehydrocholesterol series (24-CH$_3$ and 24-C$_2$H$_5$), and in the case of the 24-methyl component also the 7,22-dehydro compound. The latter (ergosterol or 24-epiergosterol on the assumption the Δ^{22} bond is *trans*) appeared to be the major component, followed by 24-ethylcholesterol. The distribution is more similar to that in the low tracheophyte, *Lycopodium complanatum*, than to that in any other living system. *L. complanatum* contains 24β-methyl-7,22-*trans*-bisdehydro-cholesterol (ergosterol), 24α-ethylcholesterol (sitosterol), 24α-ethyl-22-*trans*-dehydrocholesterol (stigmasterol), and 24β-methylcholesterol (dihydro-brassicasterol) with lesser amounts of 24-epiergosterol and 24α-methyl-cholesterol (campesterol) (Nes *et al.*, 1977a).

The fatty acids present in *C. caldarium* at pH 2–3 and 45–50° range from C$_8$ to C$_{20}$ and are dominated by palmitate (C$_{16:0}$, 44%), linoleate (C$_{18:2}$, 22%), and oleate (C$_{18:1}$, 11%) (Ikan and Seckbach, 1972; Allen *et al.*, 1970; and references cited). Odd-numbered acids, representing 5% of the total, were found in the saturated series at C_{11}, C_{13}, C_{15}, and C_{17}. With decreasing temperature the percent of unsaturated fatty acid rises (Kleinschmidt and McMahon, 1970). At 20°, for instance, 30% of the fatty acid is linolenic (18:3), while it is not present at all at 55°. Allen *et al.* (1970) found mono- and digalactosyl diglyceride, sufolipid, phosphatidyl glycerol, phosphatidyl inositol, and phosphatidyl ethanolamine to be present.

2. The Genus *Mastigocladus*

This genus of nitrogen-fixing, morphologically complex blue-green algae, also known perhaps improperly (Brock, 1978), as *Hapalosiphon,* grows up to 64° and consists of a single species, *M. laminosus,* with several "races" (Castenholz, 1969). It is found worldwide in both old and new hot waters and rapidly invades environments which become hot. Only a few years after the volcanic island of Surtsey formed off the coast of Iceland this prokaryote had colonized furmaroles in the rock, although the closest hot spring is 80 km away. It also has invaded the thermal effluents of the Savannah River Nuclear Plant. The sterols do not appear to have been studied, but Holton *et al.* (1968) examined the fatty acids of the organism grown at 49°. They range from C_{10} to at least C_{18} and are dominated by $C_{16:0}$ (54%), $C_{16:1}$ (24%), and $C_{18:1}$ (18%). No odd-numbered or polyunsaturates were detected. Palmitic acid ($C_{16:0}$) represents the only saturated acid in the mixture except for 3% of $C_{18:0}$ and 1.5% combined of $C_{10:0}$, $C_{12:0}$, and $C_{14:0}$.

3. The Genus *Synechococcus*

Synechococcus lividus is a unicellular blue-green alga belonging to the order Chroococcales, which grows as an elongate rod at temperatures as high as 74° with optimal growth at 63–67°. It is ubiquitous in hot springs of Yellowstone, obligately phototrophic, and is found elsewhere in the Western United States as well as in Japan, but appears to be absent from New Zealand and Iceland. The lipids of this species do not seem to have been studied, but Holton *et al.* (1968) have reported the fatty acids of *S. cedrorum* grown at 26–30°. The principal components of the mixture were $C_{16:0}$ (47%), $C_{16:1}$ (39%), and $C_{18:1}$ (10%). No polyunsaturates or $C_{10:0}$ were detected and only traces of odd-numbered fatty acids, and $C_{18:0}$, $C_{14:0}$, and $C_{12:0}$ amounted only to 2%. This distribution is similar to that not only in the thermophilic *Mastigocladus* (Section B-2) but also to what is found with other unicellular blue-greens such as species in the genus *Anacystis*. It is worthy of note that unlike the thermophilic, eukaryotic alga *Cyanidium caldarium,* which produces large amounts of polyunsaturates at low temperatures (Section B-1), the prokaryotic alga (*Synechococcus*) does not do so.

4. The Genus *Thermus*

Thermus aquaticus grows in a temperature range of 50–79° with an optimum at 70° and habituates hot springs as well as (but in a modified form) hot-water heaters. It is a gram-negative, nonphotosynthetic aerobe with a cell wall. The major phospholipid (80% of the lipid phosphate) of the organism contains an unsaturated, hydroxylated, undecylamine, one-glycerol-per-phosphate moiety and three fatty acids (Ray *et al.*, 1971a), but is as yet incompletely studied. *Thermus thermophilus* also has a novel lipid in which there is fatty acid amide-bound (to glucose) (Oshima and Yamakawa, 1974). The first bacterium (*T. aquaticus*) also contains smaller amounts of phosphatidylethanolamine, phosphatidylglycerol, phosphatidylinositol, cardiolipin, and phosphatidic acid. The fatty acids present are both in the normal and terminally branched series. The exact composition is temperature dependent (Ray *et al.*, 1971b). At 50° the principal acids are iso-$C_{15:0}$ (32%), iso-C_{17} (38%), and anteiso-C_{17} (22%), while at 75° the composition is dominated by iso-$C_{15:0}$ (31%), iso-$C_{16:0}$ (14%), n-$C_{16:0}$ (17%), and iso-$C_{17:0}$ (36%). Anteiso-$C_{17:0}$ is 0%. It will be seen that the place of the anteiso-$C_{17:0}$ acid at the higher temperature is taken by the two iso-$C_{16:0}$ and n-$C_{16:0}$ acids when the temperature is lowered.

5. The Genus *Bacillus*

Bacillus acidocaldarius, a spore-forming, rod-shaped, gram-positive prokaryote with a peptidoglycan cell wall, is found in acid (pH 2–5) hot springs up to 65° (Darland and Brock, 1971; Brock, 1978). Its lipids include pentacyclic triterpenes (de Rosa *et al.*, 1971a) and fatty acids in the saturated series but none in the unsaturated series. Twenty-five percent of the fatty acid mixture is composed of iso-C_{17} and anteiso-C_{17}, 5% iso-C_{16}, and traces of n-C_{14}, iso- and anteiso-C_{15}, n-C_{16}, n-C_{17}, and n-C_{18}, while 11-cyclohexyl-C_{11} and 13-cyclohexyl-C_{13} ($C_{17}H_{32}O_2$ and $C_{19}H_{36}O_2$, respectively) are found in about equal amounts as the major components (de Rosa *et al.*, 1971b). The two cyclohexyl acids were the major components (about 65%) of the fatty acid mixture in cultures grown either at 50° and pH 2 or at 70° and pH 5. In a more detailed study of the effect of pH and temperature, de Rosa *et al.* (1974a) found a curious interdependence of the two environmental factors. At lower pH, increased temperature raised the proportion of iso- and anteisoacids, but at higher pH increased temperature reversed the effect with an increase in the cyclohexylacids. The cyclohexylacids are also found in rumen bacteria but at only a 3% level (Hansen, 1967).

6. The Genus *Sulfolobus*

Sulfolobus acidocaldarius in a nonphotosynthetic, lobe-shaped bacterium that oxidizes sulfur and grows in hot, acid environments of Yellowstone National Park (Brock, 1978), Italy (de Rosa *et al.*, 1974b), and Japan (Furuya *et al.*, 1977). It can live at temperatures as high as 90°, although the optimum is 70–75°. The pH optimum is 2–5. Langworthy *et al.* (1972, 1974), Mayberry-Carson *et al.* (1974), and de Rosa *et al.* (1974b , 1975, 1976) studied the lipids of *S. acidocaldarius,* of closely related thermophilic acidophiles designated simply as MT-3 and MT-4, and of *Thermoplasma acidophilum*. The lipids of *S. acidocaldarius* and MT-3 were much the same (70% polar phospholipids, 10% glycolipids, 20% neutral lipids), with no trace of ester bonds as seen by infrared spectroscopy. Treatment of the lipid mixture with hot hydrochloric acid yielded three saturated, phosphate- and sugar-free materials with the simplest empirical formulae $C_{43}H_{42-86}O_3$. The existence of a series with varying H content indicated rings were present in some of the molecules. The alkyl chains, which proved not to be derived from fatty acids, were shown to be joined to glycerol through an ether linkage. $[2\text{-}^{14}C]$ Mevalonate labeled these chains, which could be cleaved from the glycerol with refluxing hot hydriotic acid yielding C_{40} alkanes. Through the use of mass spectroscopy, polarimetry, and other means the phosphate- and sugar-free lipid was shown to be an isopentenoidal diether of glycerol with the *sn*-2,3 or L-α,β configuration. More recently Langworthy (1977) has established that this class of lipid contains two octaisopentenoidal chains cross-linking two glycerol molecules. The eight I units in a single chain are composed of two I_4 units (each corresponding to phytane) joined tail-to-tail (not head-to-head as in lycopersene, phytoene, and carotenoids). An exchange of samples between Langworthy and de Rosa *et al.* has shown that *S. acidocaldarius,* MT-3, MT-4, and *T. acidophilum* all possess this same type of lipid, but the exact structures and the relative amounts vary with the species.

In the simplest case the structure of the I_8 chain is derived from two sets of four I units. The four I units have been polymerized in a head-to-tail manner yielding the phytanyl skeleton. This I_4-system exists as such in ether linkage with glycerol (2,3-di-*O*-phytanyl-*sn*-glycerol) in halobacteria, e.g., *Halobacter cutirubum,* which grows in extremely salty water (nearly saturated NaCl) (Kates, 1972). (For the chemical synthesis of diphytanyl ethers of phosphatidic acid and cytidine diphosphate diglyceride see Kates *et al.*, 1971.) In the case of the thermophilic acidophiles (*Sulfolobus,* etc.) the phytanyl

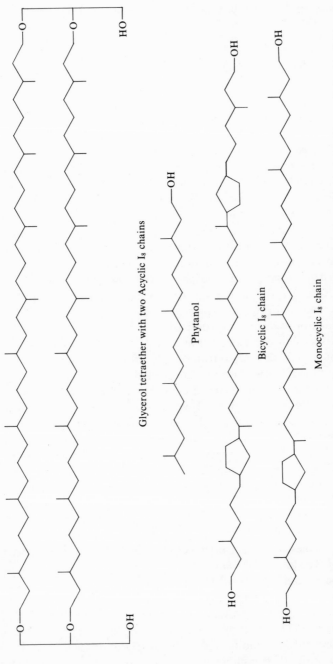

Glycerol tetraether with two Acyclic I₈ chains

Phytanol

Bicyclic I₈ chain

Monocyclic I₈ chain

Figure 12. Isopentenoid chains found variously in *Sulfolobus*, *Thermoplasma*, and *Halobacter*. (See also Chapter 7, Section B-3.)

residue has been dimerized tail-to-tail. The head (C-1) of each phytanyl piece then joins the glycerol moiety through an ether bond as shown in Figure 12. However, a modification of the polymerization of the I units also occurs. Although the I units are still joined head-to-tail (C-1 to C-4), a second condensation has occurred, leading to a five-membered ring at the juncture of the isopentenoid units. When this happens only once per I_8 piece, a monocyclic chain results. If it happens twice, a bicyclic chain is formed, and so forth. *Sulfolobus acidocaldarius,* MT-3, and MT-4 all possess large proportions (86%) of the bicyclic type, with the acyclic and monocyclic chains representing the remainder (de Rosa *et al.,* 1975, 1976).

Since the head-to-tail polymerization of I units occurs through an electrophilic attack of one unit (Δ^2-IPP, GPP, etc.) on another (Δ^3-IPP) with elimination of a proton as discussed in greater detail elsewhere (Nes, 1977; Nes and McKean, 1977), it is possible to rationalize the origin of the cyclic structures in the following manner (Figure 13). The biosynthesis of the monocycle would be initiated by the attack of geranyl pyrophosphate (GPP) on Δ^3-isopentenyl pyrophosphate (Δ^3-IPP), yielding the usual type of I_3-tertiary carbonium ion. However, instead of the usual elimination of a proton from C-2 of the incoming Δ^3-IPP to give farnesyl pyrophosphate (FPP), elimination would occur from C-3'. The intermediate triene, 3(3'),6,10-trisdehydrofarnesanyl pyrophosphate, would then undergo proton attack at C-7, producing a secondary carbonium ion at C-6 that attacks C-3' with elimination of a proton from C-2. The resulting unsaturated, monocyclic I_3 compound would then attack Δ^3-IPP in the usual manner to give the observed I_4-monocyclic skeleton. The final ether would be formed by coupling of two such I_4 units at the tail ends, reduction, and condensation to glycerol with elimination of H^+ and pyrophosphate anion.

7. *Thermoplasma acidophilum*

This prokaryotic, nonphotosynthetic, strongly acidophilic organism grows at pH 2 at an optimal temperature of 59° with a temperature limit of 65°. It forms spheres 0.3–2 μm in diameter and, particularly in young cultures, also filaments. Electron micrographs reveal no nuclear membrane or other membranous organelles in the cytoplasm. Nor is a cell wall apparent. The organism thus is akin to the mycoplasmas (Brock, 1978). It has been found in Yellowstone National Park in hot, acid soils, springs, and fumaroles as well as in burning piles of coal refuse in Indiana and Pennsylvania (Brock, 1978). Its lipids comprise phospholipids (57%), glycolipids (25%), and neutral lipids (18%) (Langworthy *et al.,* 1972). The phospho- and glycolipids contain only octaisopentenoid (instead of fatty acid) chains joining two glycerol moieties

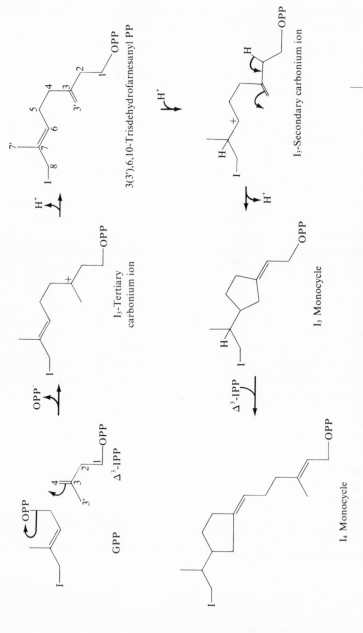

Figure 13. Proposed origin of the five-membered I₄ moiety of the isopentenoidal glycerol ethers (I = $\diagup\!\!\!\diagdown$).

through ether linkages as in *Sulfolobus*. (See Section B-6 for a deeper discussion.) The relative amounts of the various types of octaisopentenoid are, in increasing order, bicyclic, monocyclic, and acyclic. See Figure 12 for the structures.

The cytoplasmic membrane of *Thermoplasma* is about 40 Å wide according to electron micrographs. This is close to the length (about 36 Å) of the isopentenoidal tetraethers present and has led Langworthy (1977) to suggest the existence of a lipid monolayer (instead of the usual bilayer) in the membrane. Studies by Smith *et al.* (1974) indicate the membrane of *Thermoplasma* is the most rigid so far known. There is thus a correlation of sorts between the presence or absence of a cell wall in thermophilic bacteria and the kind of lipids in the membrane. Fatty acid esters (including the branched and cyclohexyl types) are found in *Thermus* and *Bacillus*, which have true cell walls, while the isopentenoidal ethers are found in *Thermoplasma*, lacking a cell wall, as well as in *Sulfolobus*, with a deficient cell wall. This correlation, however, is far from absolute and needs greater clarification, since the halobacteria contain an isopentenoidal ether, and *Thermus* and thermophilic *Bacillus* sp. contain other unusual components (Sections B-4 and B-5).

C. SUMMARY

Thermal, limnetic environments in places like Yellowstone National Park have provided excellent opportunities to discover what kinds of organism tolerate elevated temperatures. Quite an array of life forms have been found. They range from invertebrate animals and mosses (up to 50°), through less well differentiated eukaryotes, e.g., algae and fungi, which survive in places as hot at 64°, to bacteria living at temperatures of 90°. Some of the microorganisms have been the recent subjects of lipid analysis (Table 15).

Among the organisms that photosynthesize is the eukaryotic, acidophilic alga, *C. caldarium*, which will grow at 55°. It contains Δ^5 sterols with a high percentage possessing the $\Delta^{5,7}$ grouping and a 24-methyl or 24-ethyl group. Fatty acids of the usual kind dominate (C_{16-18}). Small amounts of odd-numbered chains are also present. The proportion of saturated to unsaturated—especially polyunsaturated—acids is strongly temperature-dependent. The N_2-fixing prokaryotic alga, *Mastigocladus laminosus*, allowing a temperature of 64°, also contains the usual saturated and monounsaturated fatty acids at 40°, but there are no odd-numbered or polyunsaturated chains in the mixture. The prokaryotic algae in the genus *Synechococcus*, with a temper-

Table 15. Lipids of Thermophiles

Organism	Temp. limit	Sterols	Type of alkyl chain in phospholipids, etc.	Bond to glycerol	Temp. dependent
Eukaryotic algae					
Cyanidium	60°	Yes	Normal, even and odd, sat., mono-, and polyunsat.	Ester	Yes
Prokaryotic algae					
Mastigocladus	64°	?	Normal, even, sat., and monounsat.	Ester	?
Synechococcus	74°		Normal, even, sat., and monounsat.	Ester	?
Nonphotosynthetic bacteria					
Thermus	79°	?	Nitrogen-containing as well as normal, iso, and anteiso	Ester	Yes
Bacillus	70°	(Pentacycles)	ω-Cyclohexyl C_{11} and C_{13} as well as n, iso, and anteiso, even and odd	Ester	?
Sulfolobus	90°	?	C_{40} Isopentenoid	Ether	?
Thermoplasma	65°	?	C_{40} Isopentenoid	Ether	?

ature limit of 74°, has a distribution of the common fatty acids similar to *M. laminosus* and fails to produce polyunsaturates at low temperature.

The other examined organisms are nonphotosynthetic bacteria. All have unusual lipids either partly or wholly. The aerobic genus *Thermus*, tolerating 79°, contains incompletely characterized lipids with chains containing nitrogen as well as fatty acids of the normal and terminally branched variety. The composition is markedly temperature-dependent. The acidophilic genus *Bacillus*, which will grow at 70°, contains pentacyclic triterpenes, saturated fatty acids of the normal, iso, and anteiso types, and (as the principal component) ω-cyclohexyl C_{11} and C_{13} acids. A complex but apparently not dramatic temperature effect exists. The sulfur-oxidizing bacteria in the genus *Sulfolobus* and the mycoplasma, *Thermoplasma*, which either contain no cell wall or a deficient one, possess only lipids with isopentenoidal C_{40} chains that cross-link two glycerol ethers.

The data on lipids are clearly a complex matter that seems to be related to temperature only in a marginal way. For instance, *Bacillus acidocaldarius, Thermus aquaticus,* and *Synechococcus lividus* have upper temperature limits quite close to each other from 70° to 79°, but have very different lipids. In the case of the blue-green alga, *S. lividus,* the lipid chains are of the type found in diverse organisms living at common temperatures (mesophiles), while the bacteria, *B. acidocaldarius* and *T. aquaticus,* contain in part unusual lipid chains. Viewed in reverse, the two organisms (*Sulfolobus* and *Thermoplasma*) containing the most unusual type of lipid (the ether-bonded isopentenoids) have very different temperature limits (90° and 65°, respectively). Moreover, ether-bonded isopentenoids are found in bacteria that are not even thermophilic (halobacteria, and rumen bacteria). It is interesting to note that the two organisms (*S. acidocaldarius* and *T. acidophilum*) containing only the C_{40} isopentenoids are both strong acidophiles. Unfortunately, so is the one (*B. acidocaldarius*) that contains the cyclohexyl acids and no isopentenoid ethers, and as discussed in Chapter 8, mesophilic methanogenic bacteria contain the isopentenoids. There is also no correlation as to the extent of subcellular development. Both the eukaryotic and the prokaryotic algae have common fatty acids that are similar to other algae with the respective types of subcellular development. Brock (1978) has compiled a long list of thermophiles that goes far beyond the few for which lipid data are available. When we examine this list, we find that no photosynthetic organism—eukaryotic or prokaryotic—will tolerate either acid conditions or very high temperatures. Only one blue-green alga (*Synechococcus*) and one bacterium (*Chloroflexus*) can live in water as hot as the low 70°s. In the few cases (which unfortunately do not include *Chloroflexus*) in which lipid analyses have been performed, nothing particularly unusual has been observed. All of the unusual lipids have been discovered among the nonphotosynthetic bacteria. Among these thermophilic bacteria one also finds the widest spectrum of specializations, e.g., acid requirer, hydrocarbon oxidizer, alcohol utilizer, carboxylic acid utilizer, sulfite reducer, sulfate reducer, cellulose digester, butyric acid former, methane producer, methane oxidizer, and sulfur oxidizer. Until the relationship between phylogenetics and an organism's ecological niche receives more extensive analysis, it seems difficult to say anything very profound about the relevance of thermophiles to the origin of life or to the conditions that prevailed on the primitive earth.

CHAPTER 7

Phylogenetics and Occurrence

A. ORIENTING REMARKS

The purpose of this section is to examine the extent to which there is a correlation between lipid biochemistry and organismic affinities deduced by taxonomy, in the hope of tracing familial lines (phylogenetics). Implicit in the discussion is the tentative assumption that the manner in which a cell handles lipids is a phenotypic expression of the genotype and that it is the genotype we wish to trace. Certain biochemical characteristics seem to be common to all organisms. All, for instance, seem to have ATP. This is a fairly large and complex molecule, yet apparently all cells biosynthesize it either *de novo* or from ingested parts, e.g., nicotinamide, and do not vary its structure to the exclusion of ATP itself. Such is not the case with many other molecules that as a class may be present or wholly absent or may be present but in limited and organismically varied form. These molecules would seem to be appropriate as phylogenetic markers, and many are lipids. We will first review the structural aspects and then examine biosynthesis, etc. The large amount of information now available forces us to restrict ourselves in a chapter such as this only to some of the more salient features of the subject. Instead of taking organisms one by one and summarizing their lipids, or conversely, taking molecular classes one by one and reviewing the organisms they occur in, we will ask and attempt to answer some direct questions about familial relationships.

B. AFFINITIES AMONG INDIVIDUALS AND SPECIES

1. General Remarks

When Darwin stopped over at the Galápagos Islands off the coast of Peru on his famous trip around the world, he encountered species of animals that he had not seen on the South American mainland. Among these were various finches, and he ultimately concluded (1) that they must have originated on the mainland, (2) that they had been isolated by the ocean for a long time (which modern geochemistry dates as a few million years), and (3) that the manner in which they differed from their mainland relatives indicated that change (evolution) had occurred in response to their new environment. As a result of very intensive work in the last 15 years or so, the existence of an environmental impact on the kind of animals which habituate a particular ecological niche has been extensively documented. This work has also made clear that variability at the species level extends to the level of individuals within a species. Darwin's finches themselves serve to illustrate the point. Belonging to the family Fringillidae, some of these birds are in the genus *Geospiza; G. fortis,* one of the species, inhabits several of the Galápagos Islands. However, this species in turn is constituted by a number of different types of individual (varying beak lengths, etc.), and these have different ecological niches. For instance, Grant *et al.* (1975) have recently been able to show that there is a larger variation in this species on Santa Cruz than on Daphne and that this correlates with the extent of environmental diversity on the two islands (higher on Santa Cruz than on Daphne). Generally the differences which population biologists have demonstrated at the species and subspecies levels are of the morphological (size, etc.) and behavioral (eating habits, etc.) types. These can be imagined to have their origin in relatively subtle changes of regulatory genes. Consequently, at the molecular level we should not be surprised to find some changes in the relative concentrations of lipids in various cells of the same genus or species. A more profound change would be the addition or deletion of an enzyme reflected in gross changes in the structures of the lipids present. So, we pose the question: what if any changes do occur among and within species with respect to lipids?

2. Individual Differences

Extensive measurements of the blood concentration of cholesterol in the human population have revealed substantial differences between individuals of the same sex, differences between the sexes, and differences with time (age), but there is no alteration in the kind of structures present either with the sterols or other lipids except in what are regarded as disease states. We say "what are regarded as disease" because it is not clear whether these states are fundamentally pathologic. Even though at the moment they may be quite deleterious, they may well have arisen, as appears to be the case with sickle cell anemia, in response to an environmental factor, and the new characteristic may on balance have been beneficial. In any event, these genetic diseases are instructive as to what kinds of biochemical variation can arise. The known diseases appear to be of three types: alteration in the relative concentration of enzymes, deletion of enzymes, and loss of specificity in absorption of lipids from the diet. The first case is illustrated by cerebrotendinous xanthomatosis. Mammalian sterols are composed principally of cholesterol, but its reduction product, 5α-cholestanol, is known to be a companion, and the reduction itself has been demonstrated with rat liver microsomes (Bjorkhem and Karlmar, 1974). In cerebrotendinous xanthomatosis, there is an increase in the amount of cholestanol (as well as of lanosterol, lathosterol, and cholesterol), apparently derived from an increase in the total biosynthetic rate and alterations in the relative amounts of various of the enzymes (Salen and Grundy, 1973). Similarly, in the formation of gallstones (mostly crystalline cholesterol), which, while variable in the population, has a clear familial aspect (high incidence in American Indian women), the existing evidence (Adler *et al.*, 1974) indicates not a new type of metabolism occurring but only a quantitatively different one with respect to the composition of bile (leading it to be lithogenic). The second type of variation (enzyme deletion) appears to be represented by Refsum's disease and Tay-Sachs disease (Brady, 1968), the latter having a well-established familial character (endemic in the Jewish line). Usually in human beings any phytol (derived from ingested chlorophyll) that enters systemically is degraded, but in Refsum's disease the degradation, believed to occur through α-oxidation, is impaired, and phytol becomes oxidized to phytanic acid and esterified to glycerol in the myelin sheath. This results in aberrations at the membrane level with neurological and other disorders that can be fatal. Luckily Refsum's disease can be controlled by refraining from photosynthetic plants in the diet. In Tay-Sachs disease a degradative enzyme for glycolipids, GalNAc-hexoseamidase A, is absent, and the ganglioside GM_2 accumulates with resulting neurological pathology, for which there is as yet no cure (Morell and Braun, 1972).

The third type of variation (changes in absorption) has to do with the presence of 24-alkylsterols. They are found normally in rat liver (Nes *et al.*, 1972), rat adrenals (Prost *et al.*, 1974), and cow's milk (Flanagan and Ferretti, 1974). In human pathologic states they also accumulate as in mammary (Day *et al.*, 1969; Gordon *et al.*, 1967) and brain (Paoletti *et al.*, 1971) tumors, and other lipidoses (Rao *et al.*, 1975).

Cancer in children represents a particularly saddening disease, and when it occurs in the adrenal gland, aberrations in the concentrations of the hormonal enzymes can lead to sexual abnormalities that become morphologically evident. The adrenals have the capacity to make not only corticoids but also gonadal hormones. Increases, for instance, in the mixed-function oxidase for cleavage of 17α-hydroxyprogesterone at the C-17–C-20 bond lead to testosterone and virilization of girls or prepuberal boys (pubic hair, enlargement of the clitoris or penis, etc.). Such cases again fall into our category of quantitative alterations in enzyme levels, but there is a recent case of a Japanese girl with an adrenal tumor who excretes 26- or 27-norcholesterol in her urine (Ikekawa, private communication). This desmethylcholesterol has never before been observed in the vertebrate line and represents a most fascinating type of metabolism. Sterols with altered side chains occur in marine invertebrates, and in the course of considering probable biosynthetic pathways one of us (Nes and McKean, 1977) proposed a mechanism leading from 24-dehydrocholesterol (desmosterol, which is a normal intermediate) through 26- or 27-desmethyldesmosterol to halosterol (cholesterol lacking a CH_2 group in the side chain), which was first found in the primitive chordate, *Halocynthia roretzi* (Alcaide *et al.*, 1971), and later in molluscs. The proposed 26- or 27-desmethyldesmosterol would probably serve as a substrate for the normal mammalian Δ^{24}-reductase, and the observed 26- or 27-desmethyl-cholesterol in the girl with the tumor would result. This sequence of steps requiring but two enzymes (an *S*-adenosylhomocysteine-dependent demethyl-ase and a Δ^{24} reductase), one of which (the reductase) is normal, is shown in Figure 14. It would be most instructive to know whether this desmethylsterol is really a human metabolite or is only of dietary origin. If the desmethylase is actually in the girl, it would probably mean that hidden nucleotide sequences of the sort normally expressed in more primitive organisms had been unmasked when the tumor lost some of its control at the level of regulatory genes. If this analysis could be documented, it would constitute a particularly good marker of evolution. Even if it is incorrect, the analysis given exemplifies how subspecies variation might be used to explore familial lines. The chances are that structural genes tend as a class to be more primitive than regulatory genes, and therefore the former probably offer a way to trace evolution through especially long periods of time especially when they are masked and unused in a higher organism and unmasked and expressed in the lower one. Variations in the concentration of a particular lipid between individuals or species also reflect organismic affinities, though in a less dramatic way. As we shall see in what follows, the evidence indicates that structural genes for a particular enzyme may be present in two species but expressed at different quantitative levels.

In contrast to the human population, where our knowledge of variants comes principally from disease, individual healthy plants have frequently been studied. The work accomplished so far permits a modest insight into the kinds of variation which nature does and does not allow. Unlike the fatty acids and sterols, the small isopentenoids can vary considerably within a single species. The family Lauraceae in the tracheophytic order Laurales is famous for having been the source (*Lauris noblis*) of the wreaths ("laurels") used by the ancient Greeks and Romans to crown their warriors and athletes. Many

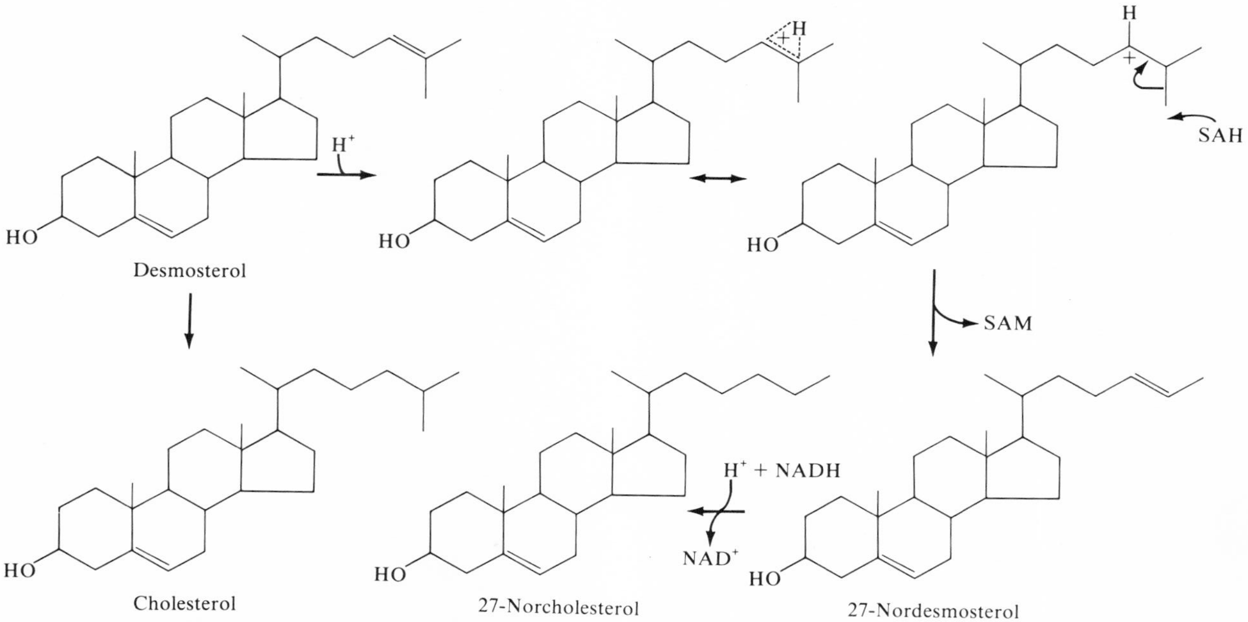

Figure 14. Possible origins of the sterol lacking one of the terminal methyl groups in adrenal carcinoma. SAH, S-adenosylhomocysteine; SAH, S-adenosylmethionine.

species are known to contain terpenes, as in the case of the genus *Cinnamomum*. The effect of habitat on the Japanese species, *C. daphnoides*, has recently been investigated by Fujita (1970, 1979). The "essential oil" consists of linalool and other components the exact composition of which varies markedly in plants collected within various parts of a single area (Kyushu District) in the vicinity of Nagasaki. Not only does the proportion of linalool vary from 16% to 86%, but as the linalool concentration drops, its place is not always taken by the same components. Fujita (1979) found four types of plants, depending on the proportions of the principal terpenes: (1) linalool (85%); (2) linalool (16–47%) and methyleugenol (23–71%); (3) linalool (16–51%), elemicine (16–64%); and (4) linalool (19–39%), elemicine (39–52%), and saffrole (7–14%). Similarly, Lincoln and Langenheim (1976) found five types of *Satureja douglasii*, which differed geographically as to whether carvone, pulegone, isomenthone, menthone, or camphene and camphor were dominant. A possible explanation for the observed divergences within a species may be that the terpenes play a pheromonal role and are not used to play a function within the plant as are the sterols and fatty acids. (For a key to the pheromonal literature, see Bowers, 1978.)

3. Interspecific Relations among Bacteria

The membranous isopentenoidal lipids of the anaerobic methanogenic bacteria (using hydrogen and carbon dioxide) show substantial interspecific homogeneity with respect to structural types, although the proportion of the individual compounds varies a lot (Tornabene and Langworthy, 1979). In the genus *Methanobacterium*, *M. thermoautotrophicum* and the mesophilic bacteria *M. ruminantium* PS, *M. ruminantium* M-1, and strains AZ and M.O.H. all possess diphytanyl glycerol diethers and dibiphytanyl diglycerol tetraethers in proportions ranging from 1 : 2 to 3 : 1, respectively, depending on the species or strain (see Chapter 6, Section B-6 for structures). The ratio in the thermophilic species is in the middle of the range. The absence of an association between the presence of these unusual lipids and the nature of the environment presumably reflects a familial rather than an adaptive relationship. Fox *et al.* (1977) have proposed a taxonomic tree for these and other methanogens, and peculiarities in the 16 S ribosomal RNA sequence of the methanogens as well as of *Thermoplasma* and *Sulfolobus* together with the similarities in their lipids (cf. Chapter 6) has led to the speculation that all of these bacteria, including *Halobacterium*, are on a very old and related evolutionary line (Woese and Fox, 1977; Fox *et al.*, 1977; Magrum *et al.*, 1978; Tornabene and Langworthy, 1979). Species variation in *Sulfolobus* is similar to what is found in *Methanobacterium* in that it is only a quantitative one, but in the former genus the range of isopentenoids is greater. *Sulfolobus* species contain not only the acyclic biphytanyl (C_{40}) structure but also mono- and bicyclic analogs (Chapter 6, Figure 12). In four cell lines the acycle varied from a trace to 30%, the monocycle from 9% to 32%, and the bicycle from 38% to 86% (de Rosa *et al.*, 1975).

In surveying the fatty acids of other bacteria, Kaneda (1977) has pointed out that among the gram-negative organisms there is a great deal of variation even at the species level. Thus, in the genus *Bacteroides*, while the fatty acids of *B. ruminicola* and *B. melaninogenicus* are mostly in the branched (presumably iso and anteiso) series, those of *B. amylophilus* and *B. succinogenes* are exclusively in the unbranched series. Similarly, the dominant acids of *Pseudomonas maltophilia* are in the iso series, but in *P. aeruginosa* and *P. cepacia* only unbranched acids exist. More homogeneity appears to be the rule within a genus, although strong variations still occur. Distinctive patterns of iso and anteiso acids (which are dominant) tend to exist within the genera *Bacillus*, *Corynebacterium*, *Listeria*, *Micrococcus*, *Norcardia*, *Propionibacterium*, *Sarcina*, and *Streptomyces*. The genus *Bacillus* is especially well studied (15 species), and Kaneda (1977) has grouped the various species into detailed categories depending on the amount of (aerobically synthesized) unsaturated acid ($<3\%$ to 28%) and on the amount, kind, and chain length of the branched fatty acid (55–95%). Most species have principally anteiso-C_{15}, and this category is called the "*B. subtilis* group." The "*B. cereus* group" has mostly iso-C_{15}, but both types have some of both acids. Despite these groupings, if an organism is grown under culture conditions in which the exogenous supply of fatty acid precursors (chain initiators) is insignificant, the fatty acid pattern becomes sufficiently individualized to serve as a fingerprint. Some *Bacillus* species also produce very unusual fatty acids. These include Δ^5 (*B. licheniformis*, *B. megaterium*, *B. pumilus*, and *B. subtilis*), Δ^8 and Δ^{10} (*B. brevis*, *B. cereus*, *B. licheniformis*, *B. marcerans*, and *B. stearothermophilus*), and cyclohexyl acids (*B. acidocaldarius*). In the last-mentioned species, the cyclohexy compounds are actually the dominant fatty acid present, and the chain length of the principal branched acids shifts to C_{17} instead of C_{15} as in the other species. The phospholipids in *Bacillus*, however, are not uncommon. They frequently are phosphatidylethanolamine, phosphatidylglycerol, and diphosphatidylglycerol, but in certain species large amounts of lysylphosphatidyl glycerol (*B. subtilis*), glucosaminylphosphatidyl glycerol (*B. megaterium*), and other basic phospholipids are formed in response to acid conditions. There is also a difference among the species of *Bacillus* in their response to amino acids in the medium. *B. cereus*, for instance, retains a predominance of iso-C_{15} when a glucose medium is replaced by one rich in amino acids, and *B. polymyxa* continues to make primarily anteiso-C_{15}; but with *B. subtilis* the dominance of anteiso-C_{15} in the glucose medium is lost in the amino acid medium through increased synthesis of the iso-C_{15} acid.

4. Sterols as a Genus Marker

Among the blue-green algae there are three cell lines for which sterols have been examined in the genus *Spirulina* (Paoletti *et al.*, 1976b). One from Mexico not clearly defined as to species contains Δ^5-24-ethyl, Δ^5-24-H, and sterols with the groupings Δ^7-24-ethyl, with lesser amounts of $\Delta^{5,22}$-24-ethyl and Δ^7-24-methyl. All the algal enzymes (including a Δ^{24} reductase) in the pathway are obviously present, though not with equal activities. *Spirulina platensis* Mao II exhibits a similar but not identical pattern (major sterols: Δ^5-24-ethyl, Δ^7-24-ethyl, and Δ^7-24-methyl). The relative activities of enzymes such as the Δ^{24} reductase and perhaps the 22-dehydrogenase apparently differ in the two species. A subspecies, Mao I, also has principally the Δ^5-24-ethylsterol, with lesser amounts of $\Delta^{5,22}$-24-ethyl, Δ^5-24-H, $\Delta^{5,22}$-24-methyl, and a Δ^5-24-ethylsterol with an additional double bond (believed to be $\Delta^{24(28)}$) in the side chain that is incompletely identified. Again, the difference from the other two cell lines seems to be quantitative rather than qualitative. For instance, the Δ^7 reductase seems to be more active in Mao I, leading to no accumulation of Δ^7 sterols.

Fifteen species or subspecies (Table 16) of the eukaryotic green algae in the genus *Chlorella* in the order Chlorococcales have been investigated (Patterson, 1969, 1971, 1974; Patterson and Krauss, 1965; Patterson *et al.*, 1974). In the several cases for which configurational information is available (Nes and McKean, 1977; Nes *et al.*, 1977a), C-24 when chiral always bears a β-alkyl group. However, variations among the species within a genus occur in the extent and position of double bonds in ring B, and in the degree of alkylation (C_1 or C_2) at C-24. The organisms can be arranged according to the biosynthetic sequence of events in ring B: Δ^8 (no examples) to Δ^7 (five examples) to $\Delta^{5,7}$ (five examples) to Δ^5 (three examples). Thus, the Δ^5 producers possess the (Δ^8 to Δ^7) isomerase, the Δ^5 dehydrogenase, and the Δ^7 reductase. The $\Delta^{5,7}$ producers lack the last enzyme, and the Δ^7 producers lack the last two enzymes. There is also a variation in the extent of alkylation. Two kinds of C_1 transferase for the first alkylation exist, one eliminating a proton from C-27, producing a $\Delta^{25(27)}$ bond, the other eliminating a proton from C-28, producing a $\Delta^{24(28)}$ bond (24-methylene). The 24-methylenesterol is then the substrate for the second C_1 transferase, also producing a $\Delta^{24(28)}$ bond, but this time at the 24-C_2 stage. $\Delta^{25(27)}$- and $\Delta^{24(28)}$-reductases are apparently present in all the species. The first C_1 transferase yielding a 24-methylenesterol must be present in six (those with 24-ethylsterols) but presumably absent in

Table 16. Species Variation in the Sterols of *Chlorella*

Species	Major sterols	
	Double bonds	Substituent at C-24
C. vulgaris	Δ^7 and $\Delta^{7,22}$	CH_3 and C_2H_5
C. emersonii	Δ^7 and $\Delta^{7,22}$	CH_3 and C_2H_5
C. fusca	Δ^7 and $\Delta^{7,22}$	CH_3 and C_2H_5
C. glucotropha	$\Delta^{7,22}$	C_2H_5
C. miniata	$\Delta^{7,22}$	C_2H_5
C. candida	$\Delta^{5,7,22}$	CH_3
C. nocturna	$\Delta^{5,7,22}$	CH_3
C. protothecoides	$\Delta^{5,7,22}$	CH_3
C. simplex	$\Delta^{5,7,22}$	CH_3
C. sorokiniana	$\Delta^{5,7,22}$	CH_3
C. saccharophila	Δ^5 and $\Delta^{5,22}$	CH_3 and C_2H_5
C. pringsheimii	Δ^5 and $\Delta^{5,22}$	CH_3 and C_2H_5
C. ellipsoidea (TU-247)	Δ^5 and $\Delta^{5,22}$	CH_3 and C_2H_5
C. ellipsoidea (TU-246)	$\Delta^{5,8}$, $\Delta^{5,8,22}$ and Δ^7	CH_3

the rest of the species (no 24-ethylsterols), and the first C_1 transferase yielding $\Delta^{25(27)}$ sterols is quite active in all but two species (those lacking 24-methylsterols). Interestingly, all of the cells that produce only sterols with two double bonds in ring B lack the first C_1 transferase, which yields the $\Delta^{24(28)}$ bond. If this were not so, 24-methylene- or 24-ethylsterols would be present, depending, respectively, on the absence or presence of the second C_1 transferase. This information suggests some sort of genetic coupling between events which lead to sterols with a given type of unsaturation in ring B and the extent of alkylation, viz., when two double bonds are present no second alkylation can occur. In the fungi this same phenomenon is observed among the Ascomycetes and Basidiomycetes in which the sterol present ($\Delta^{5,7,22}$) has only a 24-methyl group. However, the coupling is not as rigid as these data suggest, because among the green algae (division Chlorophyta) in the order Volvocales there is the organism *Chlamydomonas reinhardii*, and its sterols contain the $\Delta^{5,7}$ grouping in both the 24-ethyl- and 24-methylsterols. On the other hand, since Volvocales is believed on other grounds to be rather primitive, perhaps there is indeed a genetic coupling at the regulatory level in Chlorococcales that never developed in Volvocales.

Before leaving the sterols of green algae, we should point out that two subspecies (var. mannophila and var. communis) of *Chlorella protothecoides* have been examined and both are identical in their sterols. This information tells us that there is correspondence between sterol biosynthesis and those taxonomic parameters which lead one to believe the two types of cell belong in the same species, but not so nice a correlation exists with

Chlorella ellipsoidea. The cell lines TU-246 and TU-247 are substantially different. TU-247 has all the enzymes in the pathway, since it produces Δ^5- and $\Delta^{5,22}$-24β-ethylcholesterols [clionasterol, (65%) and poriferasterol (61%), respectively, which are the 24-epimers of sitosterol and stigmasterol] along with 24β-methylcholesterol (31%) and its 22-dehydro derivative (1%). TU-246 is strongly deficient in some of these enzymes. It produces $\Delta^{5,8}$-, $\Delta^{5,8,22}$-, and Δ^7-24β-methylsterols [ergosterol (35%), its $\Delta^{5,8}$ isomer (33%), 24β-methyl-lathosterol (17%), its Δ^8 isomer (10%), 5-dihydroergosterol (2%), and traces of 22-dihydroergosterol and two unidentified sterols]. The sterol pattern indicates at least the lack of the C_1 transferase which produces a $\Delta^{24(28)}$ sterol and a kinetic deficiency in the Δ^8 to Δ^7 isomerase and the Δ^5 dehydrogenase. Is this really a subspecies of *C. ellipsoidea*, or is it a new species? The answer depends on how one defines a species, something that still tends toward subjectivity, but clearly there is no absolute correspondence between the parameters used to classify this organism and the sterol pattern. The kinds of differences observed with TU-246 and TU-247, e.g., the lack of the $\Delta^{24(28)}$-producing enzyme, are the kinds of differences we find between cell lines assigned to different species.

In several other algal groups diversity in the sterol content between species within a genus is much less marked than with *Chlorella*, but diversity is also small within the divisions (phyla) to which the algae belong. In the Phaeophyta, for instance, regardless of the order or genus the dominant sterol is always *cis*-24-ethylidenecholesterol (fucosterol). Rarely the 24-C_1 analog (24-methylenecholesterol) and even more rarely cholesterol is found as a companion. Similarly, in the Rhodophyta either cholesterol or its 22- or 24-dehydro derivatives is nearly always the dominant sterol, and these may interchange among species. For instance, *Porphyra purpurea* (order Bangiales) principally has desmosterol (24-dehydrocholesterol), while in four other species of this genus it is replaced by cholesterol, which is more common in the division as a whole. In the order Gelidiales five species of the genus *Gelidium* all have cholesterol, but in two of them one finds an accompanying cholestanol that is quite uncommon. In the order Ceramiales four species of *Laurencia* have cholesterol as usual, but in *Laurencia undulata* it is accompanied by the Δ^7 analog (lathosterol) and by 24-methylenecholesterol. The Chrysophyta are more like the Chlorophyta in variation within the division, and as in the Chlorophyta the variation within the genus is substantial. Examples are *Ochromonas* species in the class Chrysophyceae and *Nitzschia* species in the Bacillariophyceae. Thus, while 24-methylsterols are usually dominant in the Bacillariophyceae and are so in four of the five *Nitzschia* species examined, in *N. longissima* cholesterol, present as a minor constituent in other of the species, is now dominant. In *N. ovalis* substantial amounts of 22-dehydrocholesterol make their appearance. Species of *Ochro-*

monas are well examined (Gershengorn *et al.*, 1968). *O. malhamensis* has almost only poriferasterol (98%), while in *O. danica* the amount of poriferasterol (a $\Delta^{5,55}$-24β-ethylsterol) drops to 58% and 21% of a mixture of 24β-methylsterols [brassicasterol (12%), ergosterol (3%), and 24β-methylcholesterol (6%)] make their appearance along with 24β-ethylsterols [clionasterol (9%) and 7-dehydroporiferasterol (12%)], differing from poriferasterol in the extent of dehydrogenation at C-22 and in ring B. It will be seen that the two species have very different capacities to introduce two C_1 groups.

Turning to the sterols of tracheophytes, we find a high degree of homogeneity within a genus, although occasional differences occur at other taxonomic levels. Several cases have been well investigated, but first we must briefly review the status of sterols in higher plants. The sterol pattern in most tracheophytes is commonly complicated (Nes and McKean, 1977). In the majority of cases the dominant sterol bears a 24α-ethyl group and is accompanied by 24-methylsterols of both the α and the β configuration in approximately 2:1 ratio (Nes *et al.*, 1976b) and to a lesser extent the 24-H sterol. In the Δ^5 series, this constitutes the homologous series (24-H, 24-CH_3, and 24-C_2H_5) of cholesterols. In the Δ^7 series it is the homologous lathosterols. The $\Delta^{5,7}$ series has never been encountered in gymnosperms or angiosperms. We call tracheophytes that have the homologous cholesterol series "main line" plants. Most plants investigated are of this sort, even though fluctuations exist in the relative proportions of the sterols. Among main-line plants it follows that there should not be much species variation, and the facts show this to be true. This is illustrated by *Nicotiana* species (Cheng *et al.*, 1971). *N. suaveolens* and *N. langsdorffi* are almost identical. The parts above ground are 44% sitosterol, 28% stigmasterol, 18% cholesterol, and 10% 24ξ-methylcholesterol in the latter and 46%, 26%, 21%, and 7%, respectively, in the former. Parts below ground are 55% stigmasterol, 32% 24ξ-methylcholesterol, 11% sitosterol, and 2% cholesterol in *N. langsdorffi* and 39%, 30%, 29%, and 2%, respectively, in *N. suaveolens*. Similarly, 41 species of *Brassica* in 21 genera all have sitosterol, 24-methylcholesterol, and cholesterol with some variation in minor components (Knights and Berrie, 1971). In *Brassica* seeds, 24β-methylcholesterol is frequently replaced by its 22-dehydro derivative (brassicasterol) so that only the 24α-methyl component exists with a reduced side chain. With development the 24β-methylcholesterol appears and the brassicasterol disappears. Thus, in *Brassica rapa* seeds the sterols are sitosterol (60%), stigmasterol (trace), 24α-methylcholesterol (24%), cholesterol (trace), brassicasterol (10%), 22-dihydrospinasterol (5%), and isofucosterol (trace) (Itoh *et al.*, 1973; Ingram *et al.*, 1968) with no 24β-methylcholesterol (Mulheirn, 1973), while in the leaves of mature *Brassica oleracea* leaves the sterols are sitosterol (69%), 24α-methylcholesterol (20%), 24β-methylcholesterol (10%), and cholesterol (17%) (Nes *et al.*, 1976b). Turning to plants which are not main line, we still find correspondence at the species level, regardless of other factors. Seeds, for instance, of *Citrullus vulgaris* are 58% 25(27)-dehydrochondrillasterol, 27% its 22-dihydro derivative, and 15% spinasterol, while the same sterols are present in seeds of *C. sativus* at levels of 78%, 10%, and 10%, respectively (Sucrow and Reimerdes, 1968; Sucrow and Girgensohn, 1970; Sucrow *et al.*, 1974). When we take seeds of very different families with other groups of sterols, again correspondence at the species level is found. The oils of seeds of *Camellia japonica* and *C. sasangua* are very close in composition: spinasterol (45% vs. 60%), dihydrospinasterol (45% vs. 28%), avenasterol (6% vs. 6%), methylenelathosterol (about 1%), and other (1% vs. 5%) (Itoh *et al.*, 1974); and in the genus *Clerodendrum*, the three investigated species (*C. infortunatum*, *C. campbellii*, and *C. splendens*) have identical sterols [only 25(27)-dehydroclionasterol (clerosterol) and its 22-dehydro derivative], although each of the species has primarily one or the other (Bolger *et al.*, 1970a; Bolger *et al.*, 1970b; Manzoor-I-Khuda, 1966; Nes and Pinto, unpublished observations).

The sterols of fungi (Wassef, 1977; Weete, 1974) conform to the pattern set by the red and brown algae; that is, within a larger taxon in which there is little diversity as a whole, homogeneity is reflected at the species level. For instance, all Ascomycetes and Homobasidiomycetes have dominant sterols that bear a 24-methyl group (usually ergosterol), and typing of species within this group through sterols is almost impossible due to uniformity. Thus, all wild *Saccharomyces* species and yeasts in general contain almost exclusively ergosterol when properly examined, although the amount differs drastically. *Saccharomyces* is unusual in having a very high content (Dulaney *et al.*, 1954). In six species of the Deuteromycete genus *Tricophyton* ergosterol is also found. This is not to say that all fungi contain ergosterol, which is not the case, but where several species of a given genus have been studied there is little or no variation. A particularly interesting and different case from the yeasts is comprised by the aquatic phycomycetes. In four species of *Allomyces* cholesterol is the dominant sterol (Southall *et al.*, 1977). We will examine variation above the species level in another section.

5. Interspecific Significance of Fatty Acids, Fatty Alcohols, and Hydrocarbons

The fatty acids in a number of species of *Chlorella* have been studied and found to be qualitatively but not quantitatively the same, with the major chain lengths ranging from C_{16} to C_{18} containing zero to four double bonds (Erwin, 1973; Frasinel *et al.*, 1978; Patterson, 1970; Klenk *et al.*, 1963; Chuecas and Riley, 1969; Nichols, 1965). The quantitative variations are very large, with the major fatty acid in some species being $18:2$ as in *C. vulgaris*, $18:3$ as in *C. variegata*, and saturated as in *C. sorokiniana*. The same species (*C. pyrenoidosa*) reported by different investigators also varies substantially, e.g., from 14 to 46% of $18:1$. The differences may be environmental. Patterson (1970) and MacCarthy and Patterson (1974a,b) demonstrated a complicated effect of temperature and Mg^{2+}, Ca^{2+}, and K^+ ions on the fatty acid composition of *C. sorokiniana*. Patterson (unpublished observations) has also found differences in fatty acid composition when various algae are grown heterotrophically vs. autotrophically. The fatty acid composition of three species each of *Anabena* and *Nostoc* among the blue-green algae is very much the same within each genus even at a quantitative level (variations of about 10% or less), although differences occur between the genera. However, in the genus *Anacystis*, *A. montana* and *A. cyanea* have substantial amounts of $18:2$ (10–18%) and $18:3$ (5–19%), while *A. marinus* and *A. nidulans* have no polyunsaturates at all (Erwin, 1973). Within the unicellular genus *Synechococcus*, differences of this sort among species are so strong (Kenyon, 1972) that there is very little similarity among the different cell lines except in the common absence of $18:4$, which is found in filamentous genera. For instance, cell "subgroup 1, strain No. 6708" of *Synechococcus* has $C_{14:0}$ (14%), $C_{14:1}$ (2%), $C_{16:0}$ (29%), $C_{16:1}$ (40%), $C_{18:0}$ (2%), and $C_{18:1}$ (6%), while "subgroup 6, strain No. 7003" has $C_{14:0}$ (1%), $C_{16:0}$ (29%), $C_{16:1}$ (11%), $C_{18:0}$ (2%), $C_{18:1}$ (18%), $C_{18:2}$ (18%), and $C_{18:3}$ (10%). It will be seen in the latter cell type that the C_{14} saturate has almost disappeared, the C_{16} monounsaturate has been drastically reduced, the C_{18} monounsaturate has been strongly increased, and C_{18} di- and triunsaturates make their appearance. Similar divergence is found within the unicellular genus *Aphanocapsa* and the filamentous genus *Oscillatoria* (Kenyon *et al.*, 1972).

The fatty acids of aquatic phycomycetes (Southall *et al.*, 1977) and yeasts show a degree of homogeneity, although marked species variations do exist (Kaneko *et al.*, 1976; Wassef, 1977). *Saccharomyces rosei* contains primarily $C_{16:1}$ (48%), $C_{18:1}$ (33%), and $C_{16:0}$ (15%), and these fatty acids in *S. carlsbergensis* have much the same percentages (51%, 35%, and 6%, respectively) as they do in *S. cerevisiae* (about equal amounts of $C_{16:1}$ and $C_{18:1}$ and a smaller amount of $C_{16:0}$). However, in *S. rouxii* the $C_{16:1}$ is largely replaced by $C_{18:2}$, and in *S. cerevisiae* var. *ellipsoidus* all of the fatty acids are reported to be saturated (Maurice and Baraud, 1967). *Saccharomyces cerevisiae* and two species of *Cephalosphorum* fail to alter their fatty acids significantly in response to temperature changes (Hunter and Rose, 1972; Sawicki and Pisano, 1977).

A recent analysis of the fatty acids of the muscle of 18 freshwater fish including several varieties of bass, perch, pike, and trout as well as representatives of burbot, crabbie, drum, salmon, smelt, sucker, and sunfish show a remarkably constant composition (Kinsella *et al.*, 1977). The same 16 fatty acids are present in all of them, and the proportion of any one rarely varies by more than a few percent.

The fatty acids of tracheophytes show as do their sterols a strong uniformity within a genus in the few cases studied. Examples are the genera *Myosotis* (photosynthetic) (Jamieson and Reid, 1969) and *Monotropa* (nonphotosynthetic) (Stanley and Patterson, 1977). Leaves of *Myosotis scorpioides* have mostly but not entirely straight-chain acids. Of these the main ones are $C_{14:0}$ (0.7%), $C_{16:0}$ (15%), $C_{18:0}$ (2%), $C_{16:1}$ (2%), $C_{18:1}$ (2%), $C_{18:2}$ (19%), $C_{18:3\omega6}$ (14%), $C_{18:3\omega3}$ (24%), and $C_{18:4}$ (15%) and a few percent of C_{20-24}. The respective composition of *M. arvensis* is 0.9%, 14%, 2%, 1%, 10%, 23%, 12%, 18%, and 7% and a few percent of C_{20-24}. The composition of *M. alpestris* is much the same, with small differences, as in the first two, in the unsaturated components. Such differences have been shown by Jamieson and Reid (1969) to be within seasonal variation. In particular, peaks or minima (depending on the acid) are found in August. In the nonphotosynthetic angiosperm *Monotropa*, *M. uniflora* (Indian pipe) and *M. hypopitys* ("pinesap") the fatty acid compositions are extremely close to each other: $C_{16:0}$ (23–26%), $C_{16:1}$ (0–3%), $C_{18:0}$ (2–4%), $C_{18:1}$ (9–11%), $C_{18:2}$ (49–55%), and $C_{18:3}$ (6–11%). The range of percentages encompasses both species. Homogeneity is also found with seeds. The family Cruciferae has been the subject of especially extensive fatty acid analysis. The seed oils of no less than 102 species in 53 genera were examined by Miller *et al.* (1965). Substantial differences occurred among the genera with respect to amounts as well as structures, but within a genus the kinds and amounts, both relative and absolute, of the fatty acids were usually quite close. For instance, in eight species of *Alyssum* the same acids were always found, and the maximum variation in the amount of $C_{16:0}$ in the oil was only 6–11%, of $C_{18:0}$ 2–3%, of $C_{18:1}$ 10–24%, of $C_{18:2}$ 9–24%, of $C_{18:3}$ 39–66%, of $C_{20:0}$ within 1%, and of $C_{20:1}$ within 1%. It will be seen that the major acid was always $C_{18:3}$, followed by nearly equal amounts of $C_{18:1}$ and $C_{18:2}$. Variation between genera can be greater than this. All species (four) examined in the genus *Lesguerella* contained among other acids 36–39% of hydroxy-$C_{18:2}$, 10–14% of $C_{18:3}$, 2–3% of $C_{18:2}$, 13–26% of $C_{18:1}$, 2–6% of $C_{18:0}$, and 1–6% of $C_{16:0}$. This distribution, entailing hydroxylation of the $C_{18:2}$, is qualitatively as well as quantitatively different from that in *Alyssum*, yet is consistent at an interspecific level. In other genera large amounts of $C_{22:1}$ (as in *Brassica*), $C_{20:1}$ (as in *Levenworthia*), dihydroxysaturated acids (as in *Cardimine*), etc., make their appearance. While a certain amount of overlap is found, the distributions tend to become fingerprints at the genus level.

Fatty alcohols (both primary and secondary) occur widely in living systems, free and in the form of fatty acid esters and glycerol ethers (Mahadevan, 1978). Correspondence within a genus is found qualitatively but far from quantitatively. This is illustrated in the animal kingdom by wax esters of secondary alcohols in the cuticular lipid of grasshoppers in the genus *Melanoplus* and the primary alcohols of the ester fraction of waxes in the higher plant genus *Eucalyptus*. The alcohols of *M. packardii* have chain lengths of C_{25}, C_{23}, and C_{27} in decreasing order of amount, but in *M. sanguinipes*, while odd numbers in the same range are observed, the distribution is different in detail (C_{23}, C_{25}, C_{21}, C_{27}, C_{24}, C_{22}, and C_{26} in decreasing order of amount) (Blomquist *et al.*, 1972). Similarly the primary alcohols of both *E. globulus* and *E. risdoni* are mainly even-numbered ranging from about C_{16} to C_{30}; but in the former, C_{26} (52%) is major followed by C_{28} (15%) and C_{24} (11%), while in the latter, C_{16} (27%) and C_{30} (31%) dominate followed by C_{28} (11%), C_{15} (8%), and C_{18} (6%) (Horn, 1964).

The hydrocarbons of microorganisms show a great deal of variation among cell lines otherwise believed to belong to the same genus. This can be true regardless of the photosynthetic or nonphotosynthetic nature of the cells or whether they are prokaryotic or eukaryotic. The "unsaponifiables" of blue-green algae include about 10% sterol, 40% "alcohols," 40% nonisopentenoid hydrocarbons, and 5% carotenoids about half of which is β-carotene (Paoletti *et al.*, 1976a). Of the nonisopentenoidal hydrocarbons in two lines of *Spirulina platensis* and one of a Mexican isolate of *Spirulina*, 67–84% is comprised by n-C_{17}, with the remainder ranging from C_{14} to C_{27} (normal and branched), and it includes squalene. The minor components are much the same qualitatively and quantitatively, but one of the lines (from Mexico) is distinguished by the presence (5%) of an unsaturated n-C_{17}. A fourth isolate also believed to be *S. platensis* (from the Bitter Springs sediment of archeological interest) similarly has primarily (70%) n-C_{17}, but the n-C_{15} and

Table 17. Hydrocarbons of Cell Lines of *Anacystis*

Cell line	% of Hydrocarbons							
	$C_{15:0}$	$C_{15:1}$	$C_{16:0}$	$C_{17:0}$	$C_{17:1}$	Branched C_{17}	$C_{18:0}$	Unsaturated C_{21}–C_{27}
A. nidulans[a]	21	—	5	68	5	—	5	—
A. nidulans[a]	6	—	2	90	2	—	tr.	—
A. nidulans[b]	23	—	8	44	20	—	2	—
A. cyanea[b]	—	—	—	87	—	13	—	—
A. montana[b]	—	—	—	12	—	—	—	73
A. montana[c]	tr.[d]	4	tr.	93	—	—	—	—

[a]Winters *et al.* (1969).   [c]Murray and Thomson (1977).
[b]Gelpi *et al.* (1970a).   [d]tr., trace.

n-C_{16} components, amounting only to 6% in the other three lines, are increased to 30% (Gelpi *et al.*, 1970a). Even within a "species," quantitative variation clearly can occur. In the genus *Anacystis* the discrepancies at the subspecies level are also demonstrable (Table 17). Strains of *A. nidulans* are composed of n-$C_{15:0}$, n-$C_{16:0}$, n-$C_{17:0}$, and n-$C_{17:1}$, with the major single hydrocarbon being n-C_{17} as with *Spirulina,* but the relative amounts of the hydrocarbons vary strongly, even when two strains of the same species of *Anacystis* are compared in the same laboratory. When two different species of *Anacystis* are examined, strong structural divergences appear in the form of branched hydrocarbons (7-methyl C_{17} and 8-methyl C_{17}) in *A. cyanea* and of mono- and diunsaturated hydrocarbons in the range from C_{21} to C_{27} in *A. montana* (Gelpi *et al.*, 1970a). Moreover, *A. montana* does not give the same results in two different laboratories (Table 17). Although the reasons for such discrepancies have been considered by Murray and Thomson (1977), clear explanations are not available. Bacterial hydrocarbons also show strong variations even at the subspecies level, as in the case of *Sarcina lutea,* which possesses hydrocarbons with an odd-to-even ratio in the range of 2 : 1–1 : 1. In strain FD-533, 65% of the mixture is C_{29}, 13% C_{28}, and 18% C_{27}. In strain ATCC-533 the C_{29} drops to 3%; C_{26}, C_{27}, and C_{28} are about 25% each; and C_{25} and C_{24}, amounting to only 1% in FD-533, are increased to 18% in ATCC-533 (Tornabene *et al.*, 1967; Albro and Dittmer, 1969). This large a difference incidentally, is not reflected in the fatty acids. Both strains contain predominantly (70–94%) a branched C_{15} acid, and differ only in the relative proportion of this one acid to the totality of other acids present, without much change in the relative proportions among the other acids.

Eukaryotic algae seem to display nearly as much variation in their hydrocarbons as do their prokaryotic relatives. In the genus *Chlorella* one unidentified species has primarily saturated and unsaturated C_{17} hydrocarbons [30% Δ-n-C_{17} (branched) and 25% n-C_{17}] (Paoletti *et al.*, 1976a). *C. pyrenoidosa* is similar (unsaturated C_{17} being 77% and saturated 19%) (Gelpi *et al.*, 1970a). Quite different is the report on *C. nocturna, C. ellipsoidea, C. vannielli,* and *C. vulgaris* grown autotrophically (Patterson, 1967). These species are thought to show little or no preference for chain length within the range from C_{17} to C_{36} except that chains longer than C_{29} are in progressively smaller amounts. The ratio of even to odd is reported at about 1 : 1. In addition, *C. vulgaris* but not several other species is thought to produce large amounts of unsaturated C_{25} and C_{27} hydrocarbons under heterotrophic conditions. Although the identifications in the latter work leave much to be desired, other investigators have reported quantitative differences among green algae in the genus *Uronema* and have obtained independent evidence for the presence of an unsaturated C_{27} (Paoletti *et al.*, 1976). *U. gigas* hydrocarbons are 27% iso-C_{17}, 13% n-C_{17}, 6% n-C_{25}, 10% an unsaturated C_{27}, and 17% unidentified acids, with smaller amounts of squalene (4%) and material ranging down to n-C_{14}. On the other hand, *U. terrestre* contains hydrocarbons that are 3% iso-C_{17}, 25% n-C_{17}, 10% n-C_{19}, 11% a substance thought to be an unsaturated C_{27}, 25% squalene, and lesser amounts of substances down to C_{14}.

Variability in hydrocarbons within a genus extends to the more highly differentiated level in insects, where these compounds comprise an important part (90% in cockroaches and 60–80% in grasshoppers) of the cuticular lipids (Jackson and Blomquist, 1976). They are formed as linear, branched, and unsaturated chains by an elongation–decarboxylation pathway (Major and Blomquist, 1978). Branching is believed to occur during elongation. Four species of cockroach in the genus *Periplaneta* have been well investigated (Baker *et al.*, 1963; Jackson, 1970, 1972; Major and Blomquist, 1978). Very large differences occur both in amounts and structures, and a strong species-dependent sexual influence is also found (Table 18). The major hydrocarbon of the female (13-methyl-C_{25}) is replaced by *cis*-Δ^9-C_{23} in some males. Conversely, little individual variation occurs. Similarly, the stage of development makes little difference with *P. australasiae,* still less with *P. fuliginosa,* and none at all with *P. brunnea*. Furthermore, *P. americana* differs markedly from the other three species in having as the major component a diunsaturated C_{27}, which the others lack entirely, and *P. americana* lacks 13-methyl-C_{25} as well as various hydrocarbons with fewer C atoms that are present to large extents in the others. Based on the lipid analyses it would appear that enzymes for both branching and desaturation are altered structurally from one species to another except in the pair *P. australasiae* and *P. fuliginosa,* in which the hydrocarbons are qualitatively the same. Changes in lipid structures presumably mean that changes have occurred in enzyme structures that in turn should have their origin in structural genes. When the same lipids are found but in different quantities, one could explain the changes as differences in regulatory genes. One wonders, therefore, with these cockroaches, whether we do not have three instead of four species, each with the same structural genes. *P. americana* is distinguished, for instance, by genes for the $\Delta^{6,9}$ desaturation, *P. brunnea* by no genes for desaturation, and the other two species by genes for Δ^6 desaturation that are just regulated differently within the pair. Similar arguments could be made in other cases where lipid structures differ from "species" to "species."

6. Interspecific Sesqui-, Di-, and Triterpenoids

The presence of halogen, especially iodine, in seaweed has been known for a great many years, but only recently have extensive investigations been made of its chemical form. Seaweeds are algae that have some cellular

Table 18. Hydrocarbons of the Cuticle of Cockroaches in the Genus *Periplaneta*[a]

Animal	Sex	n-C_{23}	cis-Δ^9-C_{23}	3-Methyl-C_{23}	11-Methyl-C_{23}	n-C_{25}	3-Methyl-C_{25}	13-Methyl-C_{25}	n-C_{27}	cis,cis-$\Delta^{6,9}$-C_{27}	n-C_{29}
Adults											
P. americana	—	—	—	—	—	10	16	—	—	70	—
P. fuliginosa	M	15	29	8	6	—	—	24	6	—	2
P. fuliginosa	F	14	4	13	10	—	—	45	6	—	2
P. brunnea	M	12	<0.2	15	18	—	—	46	2	—	2
P. brunnea	F	11	<0.2	15	20	—	—	48	1	—	2
P. australasiae	M	14	46	3	2	—	—	9	5	—	4
P. australasiae	F	16	2	12	3	—	—	54	4	—	3
Nymphs											
P. americana						16	13	—	1	70	—
P. fuliginosa		15	—	14	1	—	—	57	7	—	3

[a]See text for references.

differentiation and colonial character, and the halogenated compounds, which turn out to be lipids, have now been discovered both in the seaweeds *per se* as well as in some unicellular algae such as *Ochromonas*. The marine red algae (Rhodophyta) and the brown algae (Phaeophyta), both of which have representatives that are common seaweeds, have been especially well studied, but the red algae seem to be unique in their ability to incorporate three of the halogens (Cl, Br, and I) into organic compounds. Fenical (1975) has recently reviewed this subject. For our present purposes it is interesting to see whether there is interspecific homogeneity, and indeed a certain kind does exist. Thus, in the genus *Laurencia* a dozen species have been examined. This genus is characterized by having cyclic brominated and chlorinated sesqui- and diterpenes (Figure 15) presumably derived from isomers of farnesyl and geranylgeranyl pyrophosphates (FPP and GGPP).

The most usual of the sesquiterpenes has the skeleton of chamigrene in which two six-membered rings are fused in a spiro arrangement (one common C atom) at what would be C-5 in the steroids. The biosynthesis of the chamigrene type is readily envisioned as proceeding through a modification of the cyclization of squalene as shown in Figure 16. The cyclase would adsorb FPP (or more likely its *cis* or tertiary isomer) such that C-1 is adjacent to the double bond at C-6–C-7. Electrophilic attack at the terminal double bond as with squalene but in this case by oxidized bromine (Br^+) (route *a* in Figure 16) with elimination of pyrophosphate anion and abstraction of a proton from C-7′ or C-8 (FPP numbering) would give the two observed isomeric chamigrenes. In these isomers the double bond in the ring bearing the bromine atom is either endocyclic or exocyclic and presumably would have had to be formed by different enzymes with different positioning of the deprotonating agent. A third mode of cyclization (with still another enzyme) could reasonably be oxidative abstraction of hydrogen (route b in Figure 16, removal of H^-) from the acyclic precursor instead of Br^+ attack. This would lead to a triene instead of a brominated diene. All three of these bicyclic types are observed in various species of *Laurencia*. In one of the species, *L. nidifica*, the triene type and the brominated diene both occur, indicating the genus is characterized by a family of cyclases. Actually, this family is still broader, since sesquiterpenes (including a monocycle, biphenyl-type bicycles, and fused five- and six-membered bicycles) have been isolated from various species. Making the problem even more complicated, nature adds additional enzymes to oxidize and halogenate further, particularly by attack on the double bonds. Addition of BrCl, for instance, to the exocyclic triene would give nidifidiene (Figure 16); if to the brominated diene, nidificene. Addition of HOCl to the endocyclic diene would give glanduliferol. Nidificene and nidifidiene both occur in *L. nidifica,* and glanduliferol is found in *L. glandulifera*. A further modification of the pathway is seen in the use of an

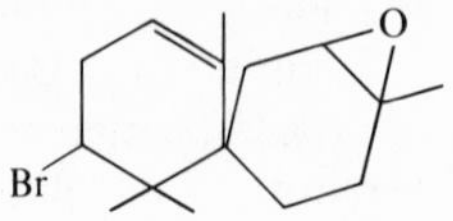

L. glandulifera 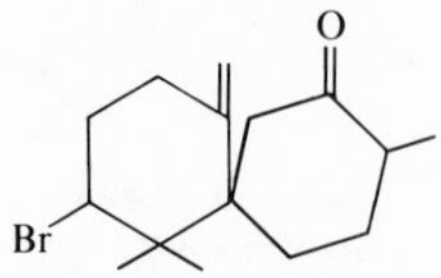   *L. glandulifera*

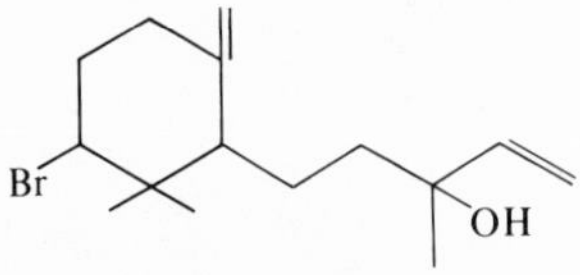

Snyderol
L. snyderae

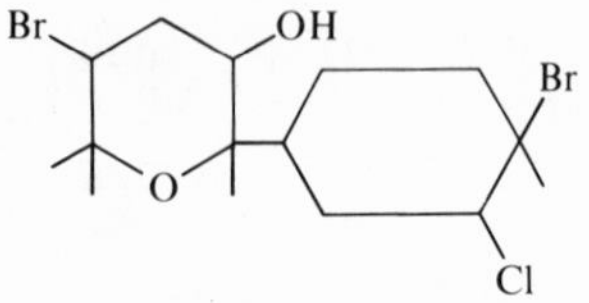

Isocaespitol
L. caespitosa

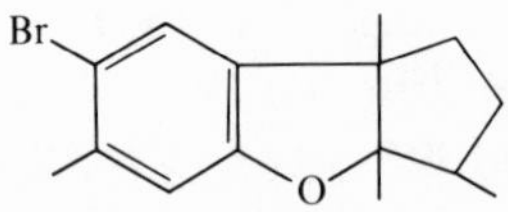

Aplysin
L. okamuri

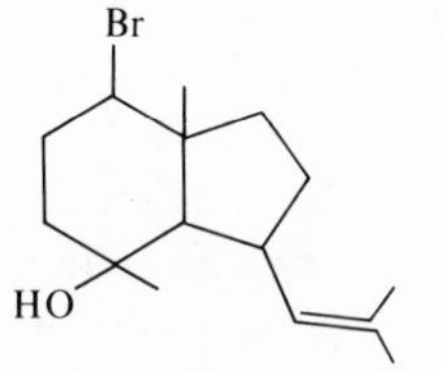

Oppositol
L. subopposita

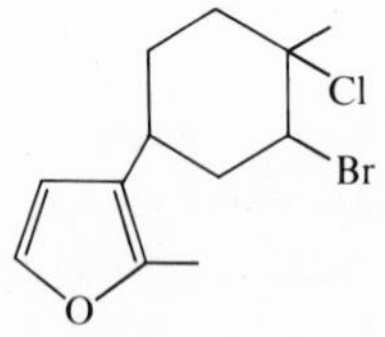

Furocaespitane
L. caespitosa

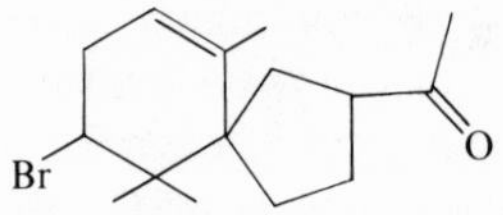

Spirolaurenone
L. glandulifera

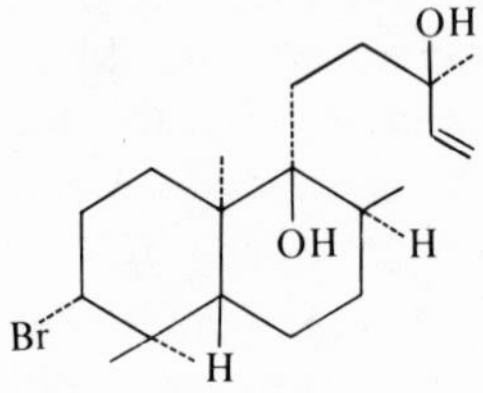

Concinndiol
L. concinna

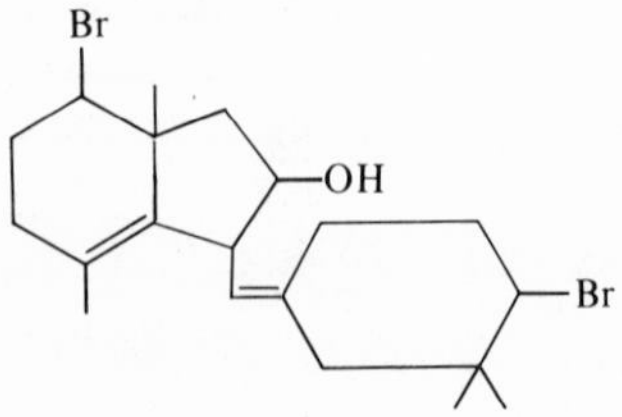

Irienol
Laurencia sp.

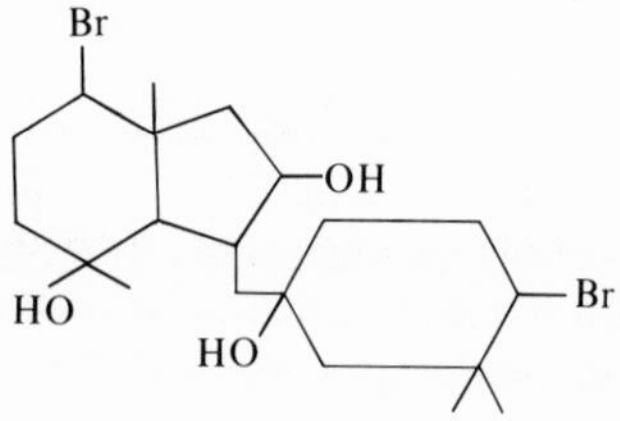

Laurencia sp.

Figure 15. Cyclic, halogenated isopentenoids of *Laurencia*. (Names for three compounds are not existent. Stereochemistry is shown only for concinndiol, which is enantiomeric in the two rings compared to steroids. For additional structures see Figure 16.)

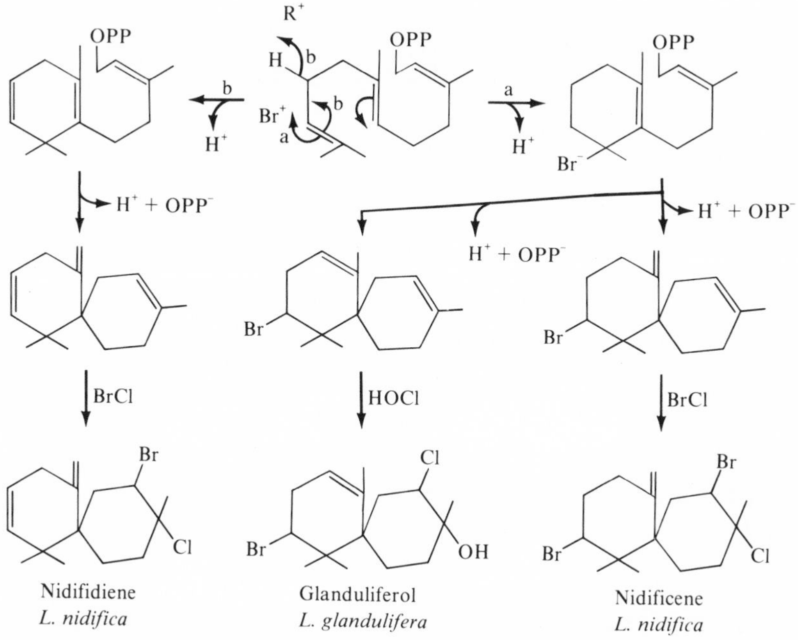

Figure 16. Probable biosynthesis of bromochamigrenes in red algae.

isopentenolog of the acyclic precursor. In this case a derivative of GGPP is cyclized to give tricyclic diterpenes, and not only are new enzymes required prior to cyclization, but the cyclases must be different from the ones leading to the cyclic sesquiterpenes.

The problem of interspecific affinities with these algae as judged by their halogenated compounds is similar to cases such as the one discussed just previously for the hydrocarbons. Different enzymes seem to be present in various species, and while there is certainly a mechanistic relationship between the enzymes, the number of types is unfortunately quite large. No less than 18 different sesquiterpenes and 6 different diterpenes have already been isolated. Frequently a given compound has been obtained from only 1 species, and only 2 species are so far known to yield the diterpenes. In the genus *Plocamium* the halogenated compounds of *P. violaceum* are cyclic mono-terpenes, whereas in *P. cartilagineum* they are acyclic, and by gas–liquid chromatography a clear fingerprint of the species results. *Laurencia* species also make halogenated acetylenes (eight-membered cyclic oxides) derived from the fatty acid pathway (Figure 17), and again different ones have been isolated from different species (Fenical, 1975). Of course, future work may reveal that all species in *Laurencia* make all of the compounds with just some being present in extremely minute quantities, but this seems unlikely. Fenical (1975) points out that 15 species of *Laurencia* synthesize at least one halogenated compound not found in the other species. In any event, something inherent in the genes of this group of organisms seems to give us a cascading effect permitting detailed modifications of the same general phenomenon, say, cyclization or halogenation, to manifest itself in different species. It is not easy to see, though, what would govern this cascading effect. Whatever the controlling factors are, some basic factor is obviously missing from our knowledge, because there is no correlation between the simple existence of isopentenoid cyclases and the cascading effect. Red algae cyclize squalene (ultimately giving cholesterol) as do more highly differentiated organisms, but in the latter, especially in the vertebrate line, we do not find such a proliferation in the number of cyclases. Nor is a proliferation of cyclases associated with halogenation, since a variety of organisms, e.g., tracheophytes, make all sorts of nonhalogenated cyclic isopentenoids (Nes and McKean, 1977) but do not halogenate them.

Although the distribution of triterpenoids has been a subject of study for decades (Boiteau *et al.*, 1964), the full, quantitative spectrum of compounds in a given plant or plant part is still not well described. Nevertheless, much can be discerned from the existing literature, and there appears to be a fairly high degree of interspecific homogeneity at least with respect to families of enzymes. A particularly extensive example of this has to do with the occurrence of the glycosides of pentacyclic triterpeniods that cause hemolysis of mammalian erythrocytes. The hemolytic activity permits quantitative analysis, and it has been applied to as many as 67 species of the genus *Primula*. Roots, leaves, stems, and flowers have each been examined separately in many cases (see Boiteau *et al.*, 1964, for references). In the majority of species,

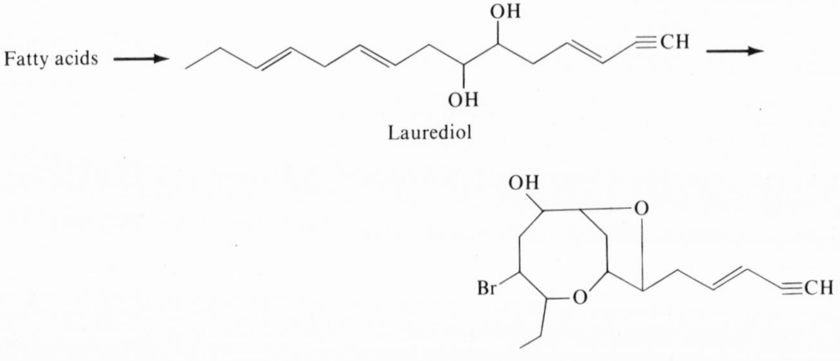

Laurediol

Laurefucin

Figure 17. Example of the halogenated cyclic acetylenes of red algae (Fenical, 1975).

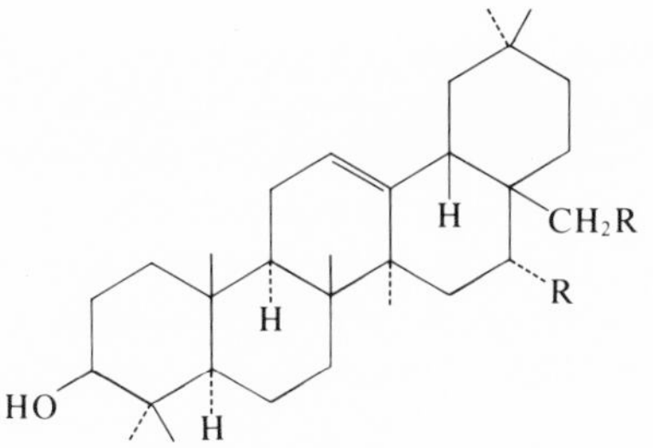

R = H, β-Amyrin
R = OH, Primulagenin-A

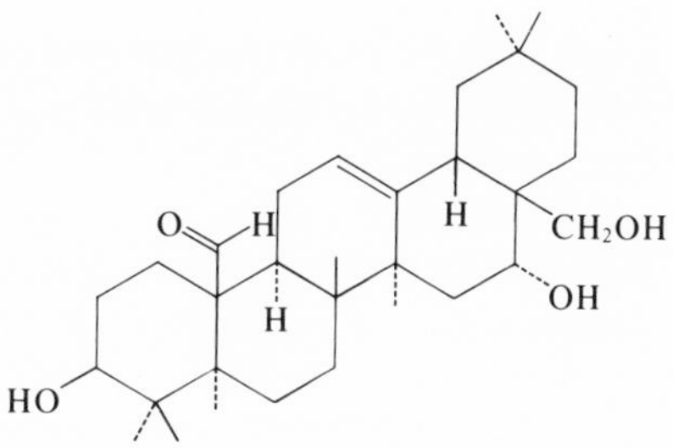

Cyclamiretine

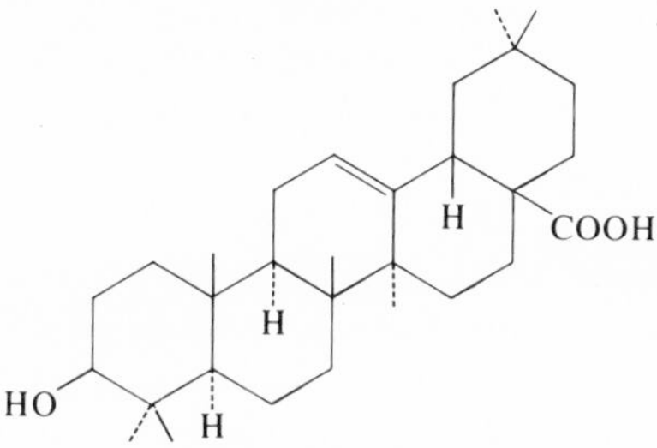

Oleanolic Acid

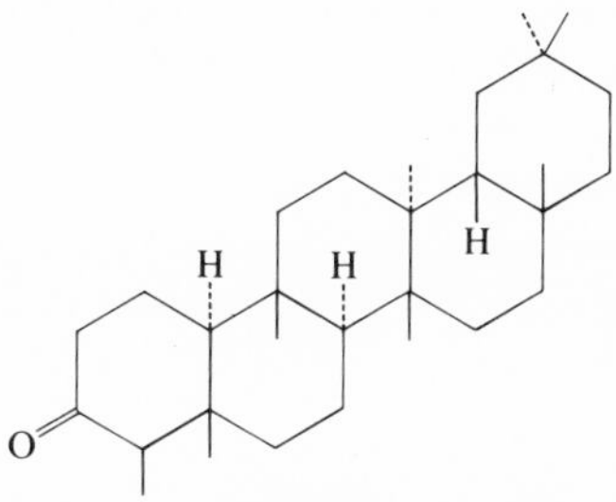

Friedelin

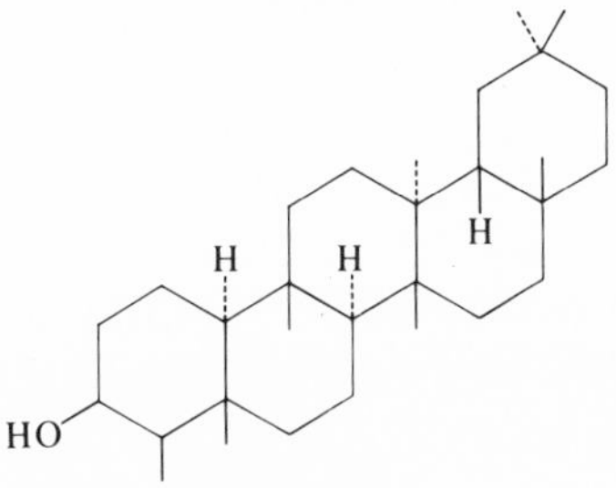

Epifriedelanol

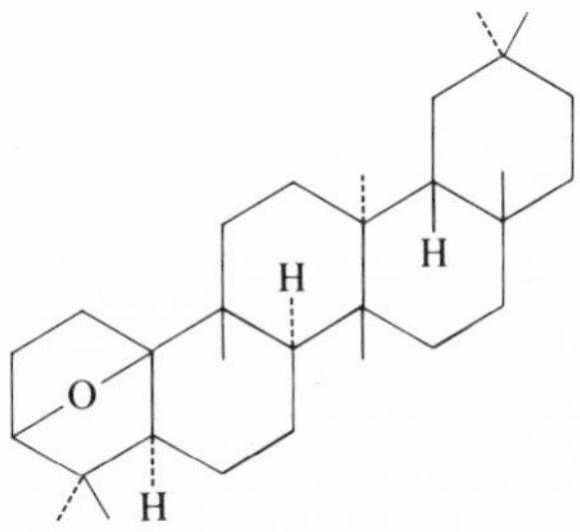

Dendropanoxide

Figure 18. β-Amyrin and friedelin series of triterpenoids.

hemolytic activity is present in all parts of the plant in quantities varying by 10^3. In only one species has no hemolytic activity been found. From its name the latter plant (*P. X. pruhoniciana* var. "Mrs. MacGillavry") appears to be a cultivar, and neither its flowers nor stems have been studied. *P. officinalis* has yielded "primulic acid" found to be a mixture of three triterpenoidal glycosides. Hydrolysis gives primarily a mole each of glucose, galactose, a uronic acid, and a triterpenoid called primulagenin-A that is a dihydroxy derivative of β-amyrin (Figure 18). Another genus (*Cyclamen*) of the family Primulaceae is also characterized by triterpenoidal glycosides, and from *C.*

europaeum has been isolated the glycoside of an aldehyde (cyclamiretine) of primulagenin-A. This aldehyde may be one of the other constituents of "primulic acid," which is known to be a carbonyl derivative of primulagenin-A.

Many pentacyclic triterpenoids also occur in the absence of glycoside formation. Some are very widely distributed. β-Amyrin, for instance, has been found in more than two dozen botanical families. In the Apocynaceae and Burseraceae this compound has been identified in 5 species of the genus *Alstonia* and in 5 species of *Canarium*, respectively. One of the carboxylic acids (oleanolic acid, Figure 18) corresponding to β-amyrin occurs in at least 33 families and often along with β-amyrin. Such is the case with the genus *Calendula* in the family Compositae. Species of Rhododendron are further interesting in that they also contain the family represented by the friedelin mode of cyclization of epoxysqualene in which C-19 (steroid numbering) has migrated to C-9 producing a positive charge at C-10. When this happens a proton can be eliminated in several ways. If it occurs from the 3β-OH group with migration of the 4β-methyl and 5α-H groups, friedelin itself results. Friedelin then by reduction of the carbonyl group leads to epifriedelanol. If the migrations do not occur, the O atom on C-3 becomes directly bonded to C-10 after elimination of the proton from the OH group, giving dendropanoxide (also known as epoxyglutinane and campanuline). *R. westlandii* is known to possess cyclases both for the β-amyrin type of cyclization (no C-19 migration) and for the friedelin type (C-19 migration occurs), since β-amyrin, friedelin, epifriedelanol, and dendropanoxide have all been isolated from its leaves. From many other species of *Rhododendron* one or both of the same two families of compounds have been isolated. Thus, the leaves of *R. metternichii* have yielded β-amyrin, ursolic acid, friedelin, and epifriedelanol, and from the leaves of *R. campanulatum* ursolic acid, epifriedelanol, and dendropanoxide have been obtained. Similar relationships are found with friedelin, friedelan-3β-ol, glutinol, β-amyrin, and taraxerol among species of the genus *Lithocarpus* (family Fagaceae) (Arthur and Ko, 1974). With tetracyclic triterpenoids among euphoids in the genus *Euphorbia* (family Euphorbiaceae) taxonomic markers also tend to be present (Nielsen *et al.,* 1979). In the latter case, euphol and tirucallol (stereoisomers of lanosterol) characterize the latex of the genus *Euphorbia* and certain other genera of the family. However, that characterizations of this sort are limited at best to the genus is illustrated by a study of Bandaranayake *et al.* (1977). In the tracheophyte family Dipterocarpaceae, comprising seven genera inhabiting southeast Asia (44 of a total of 45 species being present, for instance, in Sri Lanka), the genera *Dipterocarpus, Doona,* and *Shorea,* from 6 to 41 species of each having been examined, have the tetracyclic triterpenoid 20S-dammarenediol, but this compound is absent in 8 species of *Stemonoporus.* Furthermore, the occurrence of the pentacyclic β-amyrin is variable among the genera.

A final example of species homogeneity is found with the tetracyclic triterpenoids. In the genus *Cucumis* of the family *Cucurbitaceae*, 11 species of this genus are known to contain a highly oxygenated (tetrahydroxy-triketo) tetracyclic triterpenoidal diene acetate (cucurbitacin B) in which epoxy-squalene has been cyclized as in sterols, except that the friedelin type of C-19 migration has occurred followed by extensive hydroxylation, etc. The cyclization of epoxysqualene to give euphol (no C-19 migration) occurs frequently in the genus *Euphorbia*, although significant quantities are not present in all species.

7. Cyanolipids as Genus Markers

Recently much work, reviewed by Mikolajczak (1977), has appeared on the cyanolipids. These are fatty acid esters of branched mono- and dihydroxy C_5 nitriles. Their formally isopentenoidal structures (Figure 19) suggest they might be derived from MVA. This apparently has not been tested, but label is incorporated from L-[U-^{14}C]leucine and L-[U-^{14}C]valine (Seigler and Butterfield, 1976). The hydroxyl group of the C_5 nitrile also is found in the form of a glucoside in some members of the Sapindaceae as well as in the Leguminosae. The occurrence especially of the fatty acid esters has been extensively examined because cyanolipids I and IV (Figure 19) can be regarded as esters of the HCN addition-product (cyanohydrin) of the aldehyde with one less carbon atom. On hydrolysis of the ester link, which occurs enzymatically or by acid treatment, the addition is reversed with the liberation of HCN

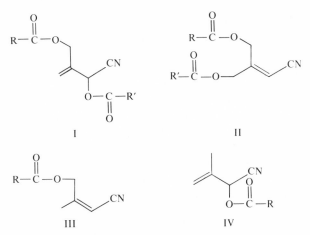

Figure 19. Cyanolipids of the Sapindaceae. R and R′ are the residues of the usual straight-chain fatty acids with total lengths of C_{14}–C_{22}. Dominant acids are variable among $C_{18:1}$, $C_{20:0}$, and $C_{20:1}$. The stereochemistry shown for III is one of two possible.

(cyanogenesis). These investigations of distribution have revealed a close correspondence among species of the same genus (Mikolajczak, 1977). Thus, the seed oil of *Cardiospermum halicacabium* is 49% I (Figure 19) and that of *C. hirsutum* is 50% I. Cyanolipid I was also detected in *C. grandiflorum* and *C. microcarpum* by NMR, which revealed I in six species of *Paullinia*. On the other hand, II is found in the oils of *Sapindus drummondi* (28%), *S. mukorossi* (13%), and *S. utilis* (32%), III in a *Stocksia* species, and IV in an *Ungnadia* species.

8. Summary

In summary, organisms that on morphological and other grounds of taxonomy have sufficiently close affinities to be called species within a genus do usually have close relationships with respect to their lipids. However, short, absolute statements about the closeness of these relationships are not easy to make, since there are various interwoven categories of variation. It would be intellectually satisfying if we could simplify the matter and say all species within a given genus have the same family of enzymes that from one species to another are varied only in amount, giving us always the same structures but in different proportions and different absolute concentrations. Actually, such often seems to be the case, but glaring exceptions exist as with the sterols of *Chlorella* and the fatty acids of *Bacillus*. In the former case certain enzymes in the biosynthetic pathway appear to be wholly missing in some species rather than just varied in amount. With *Bacillus* one of the species actually makes a type of fatty acid (the ω-cyclohexyl acids) that appears to be completely absent in the other members of the genus. This raises the difficult question as to what we mean by "genus" and "species" at a more fundamental level than classically has been applied. A definitive way of going about the classification would be to say that within a genus the structural genes are the same and that species vary not at this level but rather with respect to their regulatory genes. This would satisfy most cases we know about in which the same compounds occur but in differing proportion. Left in limbo are "problematica" such as *Chlorella, Ochromonas,* and *Bacillus*. Before dealing with them we will just have to have more information about the origin of the differences through more precise quantitation or by direct genetic examination. Perhaps all *Bacillus* species, for instance, really do possess the structural genes for the cyclohexyl acids, but if they don't, the cyclohexyl acid producers should probably be placed in a separate genus. Similar arguments can be made for the algae that produce different sterols. Despite these and other "problematica," the suggested definition of genus (structural genes the same among subtaxons) does encompass many, perhaps most cases. The

Table 19. Constant and Variant Characteristics among Species of the Same Genus in Eukaryotic Algae and Prokaryotes

Lipid	Essentially constant qualities	Main variations
Sterols	Configurations	Extent of alkylation (C_1 or C_2) at C-24, and number and position of double bonds in nucleus and side chain as well as relative amounts
Isopentenoidal membranous ethers	Precise structural types	Relative proportions of structural types—acyclic, monocyclic, bicyclic or C_{20}, C_{40}
n-Hydrocarbons, their alcohols, and acids	Range of chain lengths and proportion of even to odd numbers; exceptions include *Anacystis*	Relative proportions of chain lengths, and nature of unsaturation in the acids
Iso and anteiso acids	Range of chain lengths and dominance in *Bacillus* of C_{15} except in *B. acidocaldarius*, which has mainly C_{17}	Proportion of various acids in terms of chain length and branching as well as in relation to unbranched chains. Some species have unusual acids (Δ^5, cyclohexyl, etc.).

Table 20. Constant and Variant Characteristics among Species of the Same Genus in Animals and Higher Plants

Lipid	Essentially constant qualities	Main variations
Sterols	Configurations, extent of alkylation at C-24, position and extent of nuclear unsaturation, and relative quantities	Relative proportion of Δ^{22} and to a slight extent nuclear double bonds
Triterpenoids	Configurations, types of cyclization, and extent of oxidation	Relative proportion of compounds derived by a family of cyclizations and oxidations
n-Hydrocarbons, their alcohols and acids	Range of chain lenths, proportion of even to odd, and type of unsaturation	Detailed distribution of chain lengths, which can be substantial
Cyanolipids	Position of double bond in C_5 unit, number and position of esterified OH groups, and configuration	Amount

correlation becomes more precise the higher the ladder is climbed in cellular differentiation within an organism. Statistically, interspecific variation is encountered more often among unicelled and prokaryotic organisms than with the animals and tracheophytes. Tables 19 and 20 represent an attempt to summarize the constant and variant qualities within each lipid class as a function of cellular differentiation. In the tracyeophytes and animals, species of the same genus seem to have exactly the same enzymes with only quantitative differences (frequently remarkably small) occurring in the proportions of compounds present. Conversely, in the microorganisms, not only are much greater quantitative differences encountered but the enzymic profile appears to vary absolutely in some cases. Interspecific and inter-individual differences in lipids among tracheophytes and animals thus parallel the morphological differences found in Darwin's finches, which, for instance, all have a certain sort of beak and just vary the amount of it in three dimensions.

C. AFFINITIES AMONG GENERA AND FAMILIES

In this section we will explore the degree to which a given chemical character is distinctive for a family of genera. We shall see that a clear answer is difficult to obtain. We can begin with some of the more unusual compounds that at first might have the closest association with taxonomy.

1. Halogenated Lipids

Among the simplest halogenated substances, though not really lipids, are the bromophenols. They are broadly distributed among genera (notably *Polysiphonia, Odonthalia, Rhodomela,* and *Halopitys*) within the family Rhodomelaceae but also within other families of the red algae (division Rhodophyta). Although attempts have been made to use them as a chemotaxonomic marker, there seems to be little or no detailed correlation (Fenical, 1975). With the halogenated compounds that are true lipids the association with families is also obscure. The red algal genera *Porphyra* (Bangiales), *Gelidium* and *Pterocladia* (Gelidiales), *Gigartina* (Gigartinales), *Halosaccion* and *Rhodymenia* (Rhodymeniales), and *Spyridia* (Ceramiales) do not exhibit significant ability to halogenate lipids, while *Laurencia* and certain other genera in the order Ceramiales and *Plocamium* in the order Gigartinales do (Fenical, 1975). It follows that halogenation is not confined to a given family, since the ability is found in different orders. It is also clear that all families in a given order, say, Ceramiales, are not alike in this regard.

Reference to Section B-6 will reveal further that substantial interspecific variation occurs even within a genus when one examines the structural detail of the halogenated lipids. Halogenated diterpenes, for instance, are found only in two of the many species of the genus *Laurencia*. This restriction of structural type to particular species within a genus is confounded, however, by the opposite relationship with halogenated monoterpenes. The latter lipids are found in species as far apart as *Plocamium* and *Desmia,* which are not only in different families but in different orders (Gigartinales and Cryptonemales, respectively).

The family Rhodomelaceae in the red algal order Ceramiales is especially well investigated, and here it is satisfying to find some degree of homogeneity. Seven genera do indeed have halogenating enzymes. Unfortunately, one (*Digenia*) does not. An ecological explanation might be invoked, but it simply cannot be at the level of habitat. *Digenia* and *Laurencia,* both in the family Rhodomelaceae, inhabit the same waters, yet one does and one does not halogenate. This is not to say that habitat does not play a role. Southern California populations of *Plocamium cartilagineum* fail to have a particular trichloroaldehyde (cartilagineum), while it is present in the same species in Northern California waters.

The ability to produce halogenated lipids is obviously a complex phenomenon that transcends classic taxonomy while being related to it. One apparently has to think statistically rather than absolutely. A fair statement would be, for instance, that the probability of finding halogenated lipids in the family Rhodomelaceae is high but not 1.0. Above the family level halogenation appears to have a larger probability of occurrence in the higher evolved red algae (subclass Florideophycidae) of the orders Nemaliales, Cryptomiales, Gigartinales, Rhodymeniales, and Ceramiales than in the lower subclass (Bangiophycidae). It would be nice if we could then say that outside the red algal division the probability of finding halogenation approaches zero, but this is far from reality. Not only are the brown aglae (division Phaeophyta) well known to produce iodinated compounds, but this capacity is widespread in an enormously different group of organisms, the animals. In the vertebrates the hormone thyroxine is biosynthesized by iodination of tyrosine followed by dimerization to give the thyronine skeleton. Not only are iodinated tyrosines found in corals, protochordates, insects, molluscs, and polychete worms, but in red algae. Species in the red algal genus *Rhodymenia* (belonging to the Rhodymeniales), which have no halogenated lipids, halogenate tyrosine to give 3,5-diodotyrosine, which is a precise intermediate in the mammalian route to thyroxine. More remarkably, *Rhodymenia palmata* dimerizes as well as iodinates tyrosine, producing exactly one of the hormones (3,5,3'-triiodothyronine) that is normally bound to thyroglobulin in the mammalian thyroid gland. That iodination of tyrosine

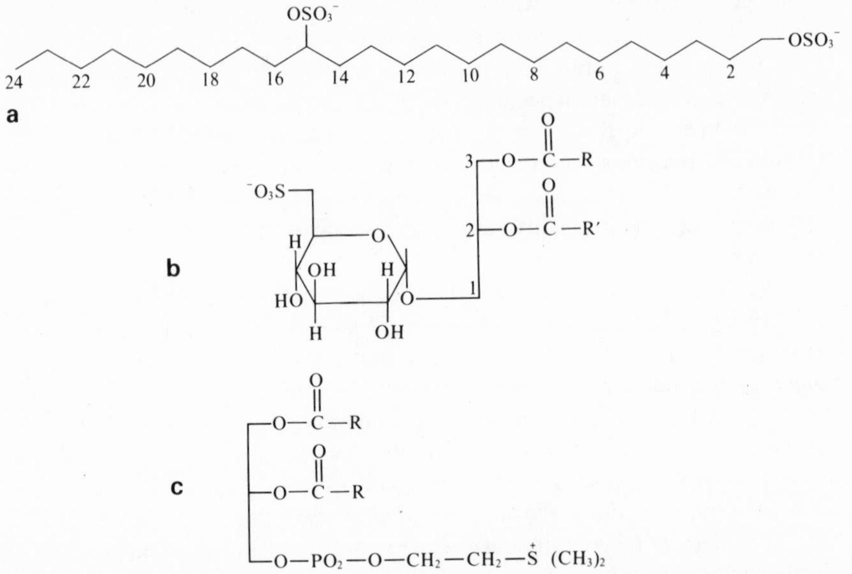

Figure 20. Structures of sulfolipids. (a) 15-Hydroxytetracosyl disulfate, a sulfatide. (Other examples exist in the C_{18}, C_{20}, C_{22}, and C_{30} series, some of which also have halogen atoms. The latter are called chlorosulfolipids. Galactosyl diglycerides in which the OH group of C-6 of the sugar has been converted to the sulfate ester is still another type of sulfatide.) (b) Sulfoquinovosyl diglyderide (a sulfonolipid). (Frequently the fatty acid at C-3 is palmitic and at C-2 linoleic. Other examples include the galactosyl analog in which the 3, 6-disulfate of glucose is bound to it through a (1 ➔ 4)-glycosidic linkage). (c) Phosphatidyl sulfocholine.

actually occurs in red algae has been demonstrated with ^{131}I in *Polysiphonia urceolata.* (Consult Fenical, 1975, for detailed references.)

Halogenated lipids are also found among other types of algae. Chlorosulfolipids (Figure 20) were discovered in the limnetic Chrysophyta by Haines *et al.* (1969) and Elvoson and Vagelos (1969) and were especially well studied in *Ochromonas* species (*O. danica* and *O. malhamensis*) (Haines, 1973). Subsequently they were found in the yellow-green Xanthophyceae (Mercer and Davies, 1974) and green Chlorophyta (Mercer and Davies, 1975). Still more recently they have been found in a wide range of classes and orders of freshwater algae but not in marine algae nor in higher plants (Mercer and Davies, 1979). The algal chlorosulfolipids (reaching values of 14% of total lipids in *O. danica* and 1.3% in the red algal genus *Porphyridium*) comprise a series of unbranched 14- or 15-hydroxy C_{22} and C_{24} fatty alcohols in which both the primary and secondary hydroxyl groups have been converted to the sulfate ester and one to five chlorine atoms have been introduced at various places. A simple example is the disulfate of 13-chloro-1,14-dihydroxy-dodecane with the R,R-threo configuration (Haines, 1979). All limnetic algae contain the C_{22} series with the possible exception of *Euglena.* Clearly, the ability to halogenate has little obvious evolutionary character. Not only are the compounds not confined to a taxonomic group, but even in the same genus (*Ochromonas*) the compounds do not have a uniform structure. *O. malhamensis* has chlorosulfolipids based on the tetracosane-1,14-diol structure. The latter C_{24} compounds are the dominant chlorosulfolipids in *O. malhamensis,* while in *O. danica* the tetracosane series is reduced to 5% and is changed to the tetracosane-1,15-diol series. The tetracosane series in terms of carbon number does, however, seem to be a marker of the genus *Ochromonas.* They have not been found in the other algae. The chlorosulfolipids are thought to replace sulfoquinovosyl diglyceride, since when the latter is high the former is low and vice versa (Mercer and Davies, 1979). Methionine supplies the S atom as well as both methyl groups of the dimethyl sulfonium moiety of phosphatidyl sulfocholine in the diatom *Nitzschia alba* (Anderson *et al.,* 1979). Evidence for the operation of such a pathway in mammalian systems also exists (Bjerve and Bremer, 1969).

2. Cyanolipids

As with the halogenated lipids, the cyanolipids (Section B-7) show a degree of homogeneity simultaneously with a vexing diversity. Seeds of plants in the family Sapindaceae frequently have these compounds (as reviewed in depth by Mikolajczak, 1977), and to some extent genera can be classified according to the type of cyanolipid present. Thus, *Cardiospermum* species have almost exclusively cyanolipid I (Figure 19), *Sapindus* species only II, *Stocksia* species almost only III, and *Ungnadia* species IV, while *Koelreuteria* species possess both II and III. However, *Paullinia, Serjania,* and *Alectryon* also are characterized by I, *Dipterodendron* by II, and *Lecaniodiscus* by II and III. As with the halogenated lipids, we have to think statistically, because there is an absolute association of cyanolipids with neither genus nor family. Quite a few genera of the Sapindaceae fail to have any cyanolipids at all. These include the genera *Deinbollia, Dodonea,* and *Maytayba*. Species studied of closely allied families (Hippocastanaceae, Melianthaceae, and Staphyleaceae) also lack cyanolipids, yet neither the ability to form materials liberting HCN nor the ability to biosynthesize a C_5 nitrile are limited to the family Sapindaceae. Cyanogenic materials are known in no less than 1000 species representing 70 families, and the formation of the hydroxylated C_5 unit containing the nitrile group is specifically known outside of the Sapindaceae. Exact overlap exists between structure and different families. *Acacia siberiana* (Leguminosae) and *Heterodendron oleaefolium* (Sapindaceae) both produce the glucoside of 2-hydroxy-3-methylbutyro-nitrile, which has a skeleton the same as cyanolipid IV (Figure 19).

It will be evident that the enzymes necessary for the biosynthesis of the relatively unusual hydroxylated C_5 nitrile are confined to neither a single genus nor a single family. Nevertheless, the presence of active enzymes for the esterification of the hydroxyl group(s) by fatty acids does appear at the moment to be a unique feature of some genera of the family Sapindaceae. There is also an association between exact structure of the nitrile and groupings of genera, e.g., compound I with *Paullinia, Serjania, Alectryon,* and *Cardiospermum.* The fatty acid composition, incidentally, shows only quantitative differences with respect to the type of cyanolipid, and there is no relationship to genus. Saturated and unsaturated chains ranging from n-C_{14} to n-C_{22} are found. Cyanolipid I from either *Cordia* or *Schleichera* has primarily $C_{18:1}$ (16–33%), $C_{20:0}$ (24–29%), and $C_{20:1}$ (22–49%). Cyanolipid II fatty acids from *Schleichera* and *Koelreuteria* are similar except for an increase in $C_{18:1}$ (24–43%) at the expense of $C_{20:0}$ (3–12%) and $C_{28:1}$ (25–55%). Fatty acids from cyanolipid III from *Koelreuteria* and *Stocksia* are mostly $C_{20:1}$ (84–86%), and those from IV are exclusively $C_{20:0}$ (28%) and $C_{20:1}$ (72%). The data suggest a very fine tuning of structure with function, i.e., a mating of the fatty acids to the nitrile for a given purpose, independent of the organism in which the structure occurs.

3. Sulfolipids

Lipids bearing a sulfur atom are an interesting group from a membranous point of view. Haines (1979) has pointed out that they are found in acidic membranes, i.e., membranes with a surface at a strongly acidic pH, e.g., 2. Under these conditions carboxylic acids will be in their un-ionized forms, but the sulfolipids with sulfate or sulfonic acid groups will exist at least partially in the anionic form preserving a charge. The exact values are calculable from the pK_as. This association with acidic membranes is independent of many parameters usually associated with taxonomy. Thus, sulfolipids are found in fungi, algae, bacteria, blue-green algae, mesophiles, acidophiles, halophiles, and the chloroplast of tracheophytes. Among the more widely distributed molecules (Figure 20) are the fatty acid 2,3-diesters of the *sn*-glyceryl 1-(6-desoxy-6-sulfono-D-glucoside). The sulfonoglucosidic moiety is described by the adjective sulfoquinovosyl, and the sulfoquinovosyl glycerols represent lipids with a sulfonic acid grouping (sulfonolipids). The two other major types are the sulfatides (lipids with a sulfate ester grouping) and lipids to which sulfocholine (dimethylsulfonium ethanol) is attached through an ester linkage. The organisms in which the sulfoquinovosyl diglycerides occur include the blue-green algal genera *Rivularia* and *Oscillatoria,* but not *Nostoc;* the photosynthetic bacterial genus *Rhodopseudomonas;* nonphotosynthetic bacteria such as the thermophilic spore former *Bacillus acidocaldarius;* some fungi among the Basidiomycetes (*Coprinus, Psalliota,* and *Clito-*

cybe); eukaryotic algae among the Chlorophyta, Chrysophyta (*Ochromonas*), Bacillariophyta, Euglenophyta, and Rhodophyta; and in higher plants. However, all of the algae in the categories given do not seem to have quinovosyl diglycerides. Thus, in the green algae *Elakatothrix viridis* and *Zygnema* sp. the quinovosyl diglyceride is replaced by sulfatides in the chlorinated alipatic series (chlorosulfolipids) as is true in a blue-green *Nostoc* species. Similarly, in the diatom *Nitzschia alba* no N-containing complex lipid is detectable (Anderson *et al.*, 1979). Similarly, the chlorosulfonolipids replace, apparently completely, the sulfoquinovosyl diglycerides in the Xanthophyta in the studied cases in the genera *Botrydium*, *Monodus*, and *Tribonema*. Partial replacement also occurs in two species of the same genus as in *Ochromonas*. *O. danica* has much chlorosulfolipid and little quinovosyl diglyceride, but the reverse situation exists in *O. malhamensis*. The Phaeophyta in the genera *Fucus* and *Pelvetia* are distinguished by having the quinovosyl diglyceride replaced not only by the chlorosulfolipids but by the galactosyl-6-sulfate analog (a sulfatide) of quinovosyl diglyceride as well as by other sulfolipids, including the disaccharide given in Figure 20. In *Halobacterium* the sulfolipids present are sulfatides in the diphytanyl glycerol ether series. (See Haines, 1979, for full structures, a list of occurrence, and references.) Sulfolipids are also present in the chloroplasts of higher plants (Davies *et al.*, 1965).

4. Fatty Alcohols

Fatty alcohols occur in both plants and animals. They are found in the free and esterified forms as well as linked through ether bonds to other molecules, especially glycerol. Mahadevan (1978) has reviewed the details of this subject, but not enough information is yet in hand to perceive taxonomic relationships or indeed if any exist. Some highlights are as follows. Human skin seems to have the most complex of the mixtures that occur. The fatty alcohols are 50% of even chains, 9% of odd chains, 26% of iso branched chains, and 13% of the anteiso variety. Monoenes show a complicated pattern among these and extend to C_{32}, while saturated alcohols are not found above C_{27} (Nicolaides, 1967). It is not obvious what one can do with this information phylogenetically, since the fatty alcohols of animals closely related to one another can be quite divergent. Some copepods have mainly $C_{16:0}$, while others yield principally alcohols with lengths of $C_{20}-C_{24}$. Similarly, the fatty-alcohol component of the wax of the honeybee (*Apis millifica*) is quite different from that of the relatively closely related bumblebee (*Bombus rufocinotus*). In the former case, largely saturated alcohols of $C_{30}-C_{32}$ are present that are mostly not in the form of esters; in the latter, monoesters are dominant, branches are important, and the chain lengths are four to six carbons shorter. It would seem that the ability to produce fatty alcohols is fairly common in biology, but not enough detail is available to perceive phylogenetic or functional generalizations. Structurally, even-numbered, unbranched chains are always dominant. In the well-studied cases, odd chains are also present. While animals could ingest these substances, making the problem more diffuse, the occurrence of both types in tracheophytes proves their biosynthesis in at least this type of organism. Dominant chain lengths are generally among C_{24}, C_{26}, C_{28}, and C_{30} in higher plants regardless of the taxonomy, although the exact distribution is not always the same. The occurrence of secondary alcohols is usually limited to the C_{29} series and to a lesser extent the C_{31} series. The hydroxyl group may be at C-7, C-10, and C-15 for the C_{29} series and C-9 in the C_{31} series.

5. The Terpenes of Photosynthetic Plants and Insects

Conifers have been known since ancient times to contain what we now recognize as isopentenoids not derived through squalene, and for several decades they have attracted chemotoxonomists. Within a given population of balsam trees in a single stand, Zavarin and Snajberk (1965) showed individual trees to have a remarkably constant composition of nine terpenes. Thus, for *Abies grandis* near Stewart's Point, California, they found santene (0.2–0.5%), tricyclene (0.8–3%), α-pinene (8–29%), camphene (17–55%), β-pinene (7–37%), Δ^3-carene (0.1–2%), myrcene (tr–0.7%), limonene (2.8–7.0%), and β-phellandrene (11–41%). The range in parentheses is for different individuals. Moreover, all species of *Abies*, regardless of locale, showed a qualitatively if not quantitatively similar pattern, and several species of another genus of the balsam (*Pseudotsuga*) did also, indicating that the balsams have a common group of enzymes for terpene formation that are varied only in amount. Quite a few additional examples of this sort of relationship within families of conifers are available, but great diversity nevertheless also exists. This is especially so within the family Cupressaceae. For instance, none of many studied species of the genera *Callitris*, *Thuja*, *Heyderia*, and *Austrocedrus* have sesquiterpenoidal hydrocarbons, while they occur but not consistently in species of *Widdringtonia*, *Cupressus*, *Chamaecyparis*, *Juniperus*, *Thujopsis*, and *Biota*. Outside of the conifers the same equivocal situation arises. The limonoids are a series of isopentenoidal polycycles starting with limonin (quite different from limonene) that can be oxidized at various places. They occur in many but not all genera of the family Rutaceae (which includes the limonoid-containing *Citrus* species). Dreyer (1966) finds that, while species within a genus seem to be homogeneous, limonoids are not confined to particular subfamilies (even though bunching occurs in two of them) but are rather evenly spread throughout the whole family at a low frequency. That the statistics of occurrence does have taxonomic meaning to some extent is indicated by the presence of limonoids in the closely related families Meliaceae and Simaroubaceae. The distribution of triterpenoids in the hopane, ursane, lupane, friedooleanane, oleanane, and arborane series in the family Fagaceae among many species of four genera also has statistical qualities. Only friedelin is found consistently. Others, such as lupeol, make their appearance sporadically (Arthur and Ko, 1974; Hui and Li, 1976, 1977).

Our understanding of phylogenetic markers with these compounds will have to incorporate in some way the statistical aspect and await better knowledge of environmental factors. Rhoades *et al.* (1976) have found that the terpenoid composition of *Satureja douglasii* (family Labiatae) is not the same when clones are grown in the laboratory and in the wild, yet they also identified two types of wild population in which strong quantitative differences in the composition of the monoterpenes occurred that were assignable to genotypic variation. The ecological factors probably associated with the terpenes are allelopathy and attraction as well as inhibition of parasitism and predation.

The literature on these substances is far too great to review here. For keys to the structure and occurrence, see Nes and McKean (1977) and Nicholas (1973) for terpenes in general, Bowers (1978) for terpenes in insects, Cordell (1974) for sesterpenes (C_{25}), Hanson (1972) for structures and occurrence of diterpenoids, Harborne (1968) and Ponsinet *et al.* (1968) for systematics of terpenoid distribution in tracheophytes, Andersen *et al.* (1977a,b) for sesquiterpenes in liverworts, Mirov *et al.* (1966 and references cited) for terpenes of Pinaceae, Pant and Rastogi (1979) for triterpene occurrence in general, and Chandler and Hooper (1979) for friedelin in particular. Terpenes with exactly the same structures as those in plants are biosynthesized by animals, notably arthropods, as in the case of α-pinene, β-pinene, limonene, and terpineol, which are ejected as cephalic secretions with diterpenes by animals in the family Termitidae and well described in pines. Rodriguez and Levin (1976) have provided a detailed review of the way terpenes connect plants with insects. Sponges also contain some of these compounds. That the enzymes for their biosynthesis exist in what are otherwise such widely divergent forms of life presents us with substantial challenges. Not only do we find overlapping in the biosynthetic patterns, but insects have obviously evolved receptor sites to distinguish terpenes made by the plants. Communication between plant and animal is not restricted to volatile components, for the tetracyclic metabolites of squalene, the cucurbitacins, formed in various tracheophytes, act to regulate the feeding habits of insects (Nielsen *et al.*, 1977). The cucurbitacins (which differ from each other in the nature of oxidation following cyclization of squalene oxide) are found in extremely divergent families of plant, e.g., the Cucurbitaceae (Lavie and Glotter, 1971), the Liliaceae (Kupchan et al., 1978), and Cruciferae (Nielsen et al., 1977), and even within a given genus their occurrence is not uniform. How the oxidosqualene cyclase (and further oxidative enzymes) arose in certain plants and not in others and how the union with the insects was established can only be construed to be a great mystery. Nevertheless, Metcalf (1979) has discussed interesting evidence for the coevolution of certain plants and insects as seen through the cucurbitacin problem. For instance, the maxillary palpa of *Diabrotica* beetles of the *Undecimpuncta* species are more sensitive to cucurbitacin B than to subsequent metabolites, suggesting an early association between the plant (squash and other cucurbits) and the insect. A related vestige of the past is also found with *D. cirstata*, *D. longicornis*, and *D. virgifera*. These insects feed on the Graminaceae, not on the Cucurbitacae. Even though the Graminaceae do not appear to contain cucurbitacins, the three *Diabrotica* insects just mentioned possess an active cucurbitacin receptor, suggesting that they arose in association with the *Cucurbitaceae* or other cucurbitacin-containing plant.

6. Bacterial Lipids

The phytanyl and biphytanyl chains bound to glycerol through an ether bond (Figure 12) are found in several genera of bacteria that have two other distinctive features (Woese et al., 1978): (1) the presence of characteristic tRNAs and ribosomal RNAs, and (2) the absence of peptidoglycan cell walls. Otherwise, they are quite diverse, including anaerobes, obligate aerobes, photosynthetic species, nonphotosynthetic species, thermophiles, acidophiles, halophiles, methanogens, sulfur oxidizers, etc. As a result of the association of the isopentenoidal lipids with the unusual RNAs and lack of a peptidoglycan cell wall, Woese et al. (1978 and references cited) have classified this group together with each other but separate from other bacteria. While this seems to be reasonable, the term "archaebacteria" that has been applied has connotations at best quite speculative. More factually, the bacteria involved are in the genera *Methanobacterium*, *Halobacterium*, *Halococcus*, *Sulfolobus*, and *Thermoplasma*. Some species, e.g., *M. thermoautotrophicum*, also contain squalene, various hydrogenated squalene derivatives, and C_{25} and C_{20} isopentenoids not bound to glycerol (Tornabene et al., 1978). The relationship these genera bear to one another in the nature of their lipids, cell envelopes, and RNAs fails to correlate with classic taxonomy (Table 21). *Sulfolobus* has been placed along with *Thiobacillus* and other gram-negative chemolithotrophs that oxidize sulfur in the family Thiobacteriaceae, order

Table 21. Incongruities of Bacteria

Order	Family	Genus	Species	Lipids
Pseudomonadales	Thiobacteriaceae	*Sulfolobus*	*acidocaldarius*	Isopentenoidal ethers
Pseudomonadales	Methanomonadaceae	*Methanobacterium*	*several*	Isopentenoidal ethers
Pseudomonadales	Pseudomonadaceae	*Pseudomonas*	*aeruginosa*	Unbranched fatty acids
Pseudomonadales	Pseudomonadaceae	*Pseudomonas*	*cepacia*	Unbranched fatty acids
Pseudomonadales	Pseudomonadaceae	*Pseudomonas*	*maltophilia*	Branched fatty acids
Pseudomonadales	Spirallaceae	*Vibrio*	sp.	Unbranched fatty acids and hydrocarbons
Mycoplasmatales	Mycoplasmataceae	*Thermoplasma*	*acidophilum*	Isopentenoidal ethers
Eubacteriales	Bacillaceae	*Bacillus*	*subtilis*	Branched fatty acids
Eubacteriales	Bacillaceae	*Bacillus*	*acidocaldarius*	Cyclohexyl fatty acids
Eubacteriales	Bacillaceae	*Clostridium*	*butyricum*	Unbranched fatty acids
Eubacteriales	Bacteroidaceae	*Bacteroides*	*ruminicola*	Branched fatty acids
Eubacteriales	Bacteroidaceae	*Bacteroides*	*amylophilus*	Unbranched fatty acids
Eubacteriales	Enterobacteriaceae	*Escherichia*	*coli*	Unbranched fatty acids
Eubacteriales	Micrococcaceae	*Sarcina*	*lutea*	Branched fatty acids and hydrocarbons
Eubacteriales	Halobacteriaceae	*Halobacterium*	*halobium*	Isopentenoidal ethers
Eubacteriales	Halobacteriaceae	*Halococcus*	sp.	Isopentenoidal ethers

Pseudomonadales. *Thermoplasma*, a mycoplasma (Brock, 1978), should be in the family Mycoplasmataceae, order Mycoplasmatales, and *Methanobacterium* presumably should be in the family Methanomonadaceae, order Pseudomonadales. It is not clear to us what the taxonomy of the halobacteria is, but their general characteristics are certainly such as to place them in families different from the others. A lack of correlation with the usual taxonomic parameters is not confined to these bacteria. As discussed below, less exotic organisms also are anomalous.

The ability of bacteria to synthesize squalene and sterols seems to have little relationship to such parameters as the character of other lipids whether or not oxygen is utilized. Squalene arises in the apparent absence of sterol in bacteria as diverse as the aerobic *Halobacterium cutirubum* (Tornabene, 1978, and references cited), *Staphylococcus* sp. (Micrococcaceae, Eubacteriales) (Suzue *et al.*, 1968), *Rhodomicrobium vanielli* (Han and Calvin, 1969), and *Rhodospirillum rubrum* (Han and Calvin, 1969). More recently squalene has been found and its biosynthesis demonstrated in nine species of *Actinomyces* and one species each of *Corynebacterium*, *Propionbacterium*, *Rothia*, *Bacillus*, *Bacterionema*, and *Streptococcus*, but not in other species of *Corynebacterium*, *Arachnia*, *Actinomyces*, and *Bacillus* (Amdur *et al.*, 1978). Neither squalene nor sterol is found in *Mycobacterium bovis* (Mycobacteriaceae, Actinomycetales) or *Aerobacter cloacae* (Enterobacteriaceae, tribe Eschenchieae, Eubacteriales) (Schubert *et al.*, 1968). Both squalene and sterols are found in the aerobic, methane-oxidizing *Methylococcus capsulatus* (Bird *et al.*, 1971; Bouvier *et al.*, 1976) as well as in *M. organophilum* (Patt and Hanson, 1978), and sterol alone but presumably from squalene is found in *Azotobacter chroococcum* (Azotobacteraceae, Eubacteriales) (Schubert *et al.*, 1968).

Kaneda (1977) has recently considered the kinds of fatty acids biosynthesized by bacteria. He has subdivided these organisms into three groups. Type I bacteria make branched-chain fatty acids. They also usually but not always have no unsaturated acid and have no oxygen requirement. Representative genera are *Bacillus*, e.g., *B. subtilis*, and *Bacteroides*, e.g., *B. ruminicola*. Type 2 make unbranched fatty acids, commonly desaturates (anaerobically), and have no oxygen requirement. *Clostridium* and *Escherichia*, e.g., *C. butyricum* and *E. coli*, are examples of this category. Type 3, which include *Pseudomonas*, e.g., *P. aeruginosa*, also synthesize unbranched fatty acids but are different from the other types in that oxygen is required and desaturation of fatty acids occurs by the aerobic route. However, as Kaneda (1977) realizes, there is a great deal of diversity in the bacterial fatty acids. The occurrence of acids with branched chains correlates absolutely with neither genus nor higher taxonomic parameters. Thus, within the genus *Pseudomonas* some species such as *P. maltophilia* have predominantly branched acids, whereas in *P. aeruginosa* and *P. cepacia* only unbranched chains are found. Within the genus *Bacteroides* one finds the same sort of disparity, and while branched chains were thought to be restricted to genera comprising the gram-positive groups of bacteria, it is now known that some gram-negative species also have such acids. In addition, there are the bacteria that biosynthesize centrally methylated fatty acids by C_1 transfer, representing an entirely different biosynthetic pattern from that utilizing branched initiators, which yields the more common branching at the end of the chain as well as those that make cyclohexyl fatty acids.

Hydrocarbons of bacteria are derived from fatty acids and seem to reflect the general type of fatty acid pattern of biosynthesis as to whether branching occurs or not. *S. lutea*, for instance, has mostly branched fatty acids (C_{15} alone amounting to 70–94%, depending on the strain), and its hydrocarbons seem also to be branched (Albro and Dittmer, 1970). Conversely, *Vibrio marinus* has predominantly the n-C_{14}, n-C_{16}, and n-C_{18} fatty acids (both saturated and unsaturated) and its hydrocarbon is principally n-C_{17} (80%) (Oro *et al.*, 1967). The relationships between the chain lengths of the fatty acids and hydrocarbons and the manner in which one is converted to the other, however, is not simple. *V. marinus* has three dominant fatty acids and only one dominant hydrocarbon. The latter (C_{17}) appears to be derived by decarboxylation of one (C_{18}) of the former. *S. lutea* exhibits a very different pattern. The hydrocarbons range from C_{23} to C_{30}, with strong differences in the detailed distribution, depending on the strain. The range of hydrocarbons is about twice that of the fatty acids (C_{12}–C_{17}). Since the hydrocarbons are branched at both ends, some sort of dimerization of the branched fatty acids is thought to occur.

The data on the fatty acids and hydrocarbons are brought together in summary form with that on the isopentenoidal ethers in Table 21. It will be evident that very little correlation exists. A given major family of enzymes, such as those dealing with the formation of isopentenoidal ethers or with the use of branched initiators in fatty acid synthesis, are confined neither to taxonomic family nor to order. Conversely, within a given order, family, or even genus the lipids can be structurally different. Particularly clear examples are: (1) the eubacterial family Bacillaceae includes species with cyclohexyl fatty acids, species with unbranched fatty acids, and species with branched fatty acids; (2) the pseudomonads include more than one family with isopentenoidal ethers as well as one family with species differing in the existence of branching and the lack of it in the fatty acids; (3) the occurrence of branched fatty acids and hydrocarbons in eubacteria and pseudomonada; and (4) the occurrence of both unbranched fatty acids and hydrocarbons in eubacteria.

Kaneda (1977) has suggested an evolution of the following sort. From an "abiotic system" arose the "branched acid system," which led to the "anaerobic system," and then to the "polyenoic system." On the other hand, Fox *et al.* (1977) and Woese *et al.* (1978) regard the methanogenic bacteria as the oldest line of life, which "may well be older than the blue-green algal one, which fossil evidence suggests to be close to 3 billion years." These authors go on to suggest the existence of two lines of bacterial evolution, viz., the eubacterial line and the one represented by the "archaebacteria." A third line is suggested to be through a "protoeukaryote" to the "cytoplasmic aspect" of the eukaryotic cell. These various speculations are based on eight main assumptions: (1) life began from nonlife on this planet; (2) the atmosphere, while containing CO_2 to accommodate the methanogens, was nevertheless reduced; (3) the 16S ribosomal RNAs constitute a phylogenetic marker; (4) the lipids constitute a phylogenetic marker; (5) a requirement for oxygen represents an advanced state (cf. point 2); (6) within the advanced group the presence of an anaerobic desaturation of fatty acids is a primitive character; (7) the presence of an aerobic desaturation is an advanced phenomenon; and (8) the presence of or lack of a peptidoglycan cell wall is a phylogenetic marker. In addition to this list of assumptions being rather long, several inconsistencies exist. They include the problem of assuming a reduced atmosphere high in hydrogen (Fox *et al.*, 1977) yet containing carbon in its most highly oxidized state (CO_2); the lack of correlation between the absence of a peptidoglycan cell wall and lipids (mycoplasmas being known with the usual fatty acids in their cytoplasmic membrane and obligately requiring sterol); and the lack of correlation, inherent in classic taxonomy, between lipids and parameters, all of which at some point have a biochemical basis. Before we can really decide something definite about the origin(s) of bacteria (and of life) we need to have an unequivocal way to decide on what shall be regarded as a phylogenetic marker (sulfur oxidation, or isopentenoidal membranes, or photosynthesis, or what?), and that really is not confidently in hand. In nautical terms we need a lighthouse that has a precise, constant set of geometric coordinates. Such lighthouses, hopefully existing somewhere, most probably lie in phenomena below the functional level, like some sorts of genetic control (perhaps the 16 S RNA) and some sorts of enzymes in biosynthetic pathways that do not influence the structure of the product. If we tentatively accept this, then it becomes at least possible to make groupings of organisms that are probably on a common phylogenetic line. When and if the lines ever diverged or had multiple origins are other questions. This argument will be resumed in a later section.

7. Lipids of Photosynthetic Prokaryotes

The photosynthetic bacteria are now classified largely in a biochemical way within the order Rhodospiralles (Buchanan and Gibbons, 1974). There are three families, depending on the kind of chlorophylls and carotenoids present. The Rhodospirillaceae and the Chromatiaceae contain bacteriochlorophylls *a* or *b*, and the Chlorobiaceae contain bacteriochlorophylls *c*, *d*, or *e* with some *a* (Gloe *et al.*, 1975). The lipid side chain of the bacteriochlorophylls is phytol in *b*, phytol along with geranylgeraniol in *a*, and farnesol in *d* and *e*. (See Gloe *et al.*, 1975, for leading references.) Very recently the thermophilic, gliding, filamentous, phototrophic *Chloroflexus aurantiacus* was shown to have still another bacteriochlorophyll, designated C_s, which has a nonisopentenoidal side chain, viz., the n-C_{18} alcohol corresponding to stearic acid called "stearylalcohol" (Gloe and Risch, 1978). A small amount of bacteriochlorophyll *a* was also present. The Chlorobiaceae species *Rhodopseudomonas spheroides*, a purple nonsulfur species, and *Chromatium vinosum*, a purple sulfur species, have been found to contain only phytol in agreement with what has just been described, and neither contained sterol (Nes, Frasinel, and Adler, unpublished observations). The phospholipids of four species of *Rhodopseudomonas* (*capsulata*, *palustris*, *spheroides*, and *gelatinosa*) and one of *Rhodospirillum* (*rubrum*) have been examined (Wood *et al.*, 1965). Phosphatidyl glycerol and phosphatidyl ethanolamine were common to all cell lines, but phosphatidyl choline, cardiolipin, sulphoquinovosyl diglyceride, and an unidentified component (perhaps *O*-ornithyl phosphatidyl glycerol) were present in various species. Whether grown aerobically in the dark or anaerobically in the light, the major fatty acids were always *n*-hexadecanoic, 9-hexadecenoic, and 11-octadecenoic acids. Small amounts of the Δ^9 isomer of the latter $C_{18:1}$ acid were also present in one species from each genus (*Rps. gelatinosa* and *R. rubrum*). From *R. rubrum* have been isolated the methyl esters of 2-methylpentadecanoic acid, 2-methylhexadecanoic acid, and 2-methylheptadecanoic acid (Maudinas and Villoutreix, 1977). By contrast, no fatty acid methyl esters were found by the latter authors in *Rhodopseudomonas spheroides*, although fatty acid methyl esters have been isolated from corn pollen and fungi. In the eukaryotic green algae and chloroplasts of higher plants, galactosyl diglycerides, sulfoquinovosyl diglyceride, and phosphatidyl glycerol are the major components and polyunsaturated acids are present (see Wood *et al.*, 1965, for leading references). The lipid side chain of eukaryotic chlorophylls is always phytol from brown algae to tracheophytes.

Blue-green algae appear to be more closely related to higher plants in their lipids than to the photosynthetic bacteria. The blue-greens have a substantial lipid content (4–9% of the dry weight of various species of *Anabena*, *Agmenellum*, *Microcoleus*, and *Spirulina*), and cytologic evidence indicates lipid "deposits" (Wolk, 1973). Among the well-characterized lipids are fatty acids, hydrocarbons, and sterols. The fatty acids are nearly always of the unbranched variety with chain lengths varying from C_{10} to C_{18}. For instance, the principal components in *Synechococcus*, *Nostoc*, *Chlorogloea*, *Oscillatoria*, and *Hapalosiphon* grown autotrophically are $C_{16:0}$ (29–54%), $C_{16:1}$ (palmitoleate, 15–39%), $C_{18:1}$ (oleate, 7–26%), $C_{18:2}$ (0–13%), and $C_{18:3}$

(α-linolenic acid, 0–21%). What consequential variation there is lies in the amounts of $C_{18:2}$ and $C_{18:3}$, which are present only in *Nostoc, Chlorogloea,* and *Oscillatoria* (Holton *et al.*, 1968). No relationship to family is obvious in the presence of the di- and triunsaturated acids. *Nostoc* and *Oscillatoria* belong to two different families, Nostococeae and Oscillatoriaceae, respectively. Holton *et al.* (1968), who can be consulted for further references as can Wolk (1973), note that the fatty acid composition may relate to the degree of morphological complexity. Unicellular algae without complexity (*Anacystis* and *Synechococcus*) lack polyunsaturated fatty acids, while the colonial marine algae (*Coccochloris* and *Agmenellum*) and filamentous species in the genera *Oscillatoria, Lyngbya, Microcoleus, Trichodesmium, Chlorogloea, Anabena,* and *Nostoc* all have polyunsaturated acids. The exception to the rule is the complex *Hapalosiphon,* which should but does not contain either di- or triunsaturated acids. Unlike the other genera, however, and for reasons which are not clear, about 50% of the fatty acids of *Trichodesmium* is composed of $C_{10:0}$ (capric acid).

Shaw (1966) and Holton *et al.* (1968) believe that the differences in the fatty acid compositions of the blue-green algae and the photosynthetic bacteria (which contain no polyunsaturates and no oleate, but do contain the otherwise unusual Δ^{11}-C_{18} compound, i.e., vaccenic acid) are too large to permit the assumption of a common phylogenesis for these two types of photosynthetic prokaryote. Since a change in the environmental conditions of organisms can induce alterations in the lipids, it is worth pointing out that Holton *et al.* (1968) investigated the lipids for autotrophic vs. heterotrophic growth of *Chlorogloea fritschii.* They observed a marked difference (strongly lowered $C_{18:3}$ with an increase in $C_{18:1}$) under heterotrophic conditions without a change in $C_{18:2}$, but the changes are not of a kind to alter the conclusions about morphological correlations or the lack of relationship between the blue-greens and the bacteria.

As a first approximation, the blue-greens contain saturated, odd-numbered, unbranched hydrocarbons, but even-numbered components, especially n-C_{16}, and unsaturates, especially n-$C_{17:1}$, have significant concentrations. Chains with a methyl branch at C-7 or C-8 are also present in some cases. Substantial interspecific variation occurs (Section B-5), making correlation at higher taxonomic levels difficult.

The sterols of blue-green algae have been reviewed by Nes and McKean (1977), who can be consulted for detailed references. (See also Section B-4.) Families of genera are not characterized by the same sterols. For instance, *Anabena* and *Nostoc* both belong to the Nostococeae in the order Nostocales, but the sterols of *A. cylindrica* have the Δ^5 and $\Delta^{5,22}$ types of unsaturation and no alkylation at C_{24} beyond the C_1 stage, while *Nostoc commune* has the Δ^7 and $\Delta^{7,22}$ types of unsaturation and the alkylation at C-24 includes both C_1 and

C_2 with the latter dominant. Similarly, in the family *Oscillatoria* (Nostocales) the primary sterols of *Phormidium luridum* have a 24-ethyl group with Δ^7, $\Delta^{7,22}$, and $\Delta^{5,7,22}$ types of unsaturation, while in species of *Spirulina*, also family Oscillatoria, both 24-methyl- and 24-ethylsterols are prominent and the pathway is completed to the Δ^5 stage, the principal sterol in all cases of *Spirulina* being 24ξ-ethylcholesterol. If we can take *P. luridum* as representative of blue-green algae with respect to photosynthesis, it is interesting to note that phytol is found in large quantities after saponification of an acetone extract (de Souza and Nes, 1969). Presumably this lipid is derived from chlorophyll as it is in higher plants. We have also found phytol in eukaryotic brown algae (de Souza and Nes, 1969).

8. Lipids of Fungi

This subject has been reviewed recently in depth by Wassef (1977) and Nes and McKean (1977). Fungi are currently grouped into the Phycomycetes, which are believed to be the least well developed, through the Ascomycetes, which include the yeasts, and Deuteromycetes (lacking sexual character; "imperfect fungi") to the Basidiomycetes, which include the mushrooms and are thought to represent the most well developed of the fungi. In addition, there are the slime molds. The major fatty acids of fungi, regardless of taxonomy, are unbranched and are usually composed mainly of the $C_{16:0}$, $C_{18:1}$, and $C_{18:2}$ acids, according to Wassef (1977). The range is from C_{14} to C_{22}, but the exact composition is quite variable. Basidiomycetes, beyond not having any $(\omega6)$-$C_{18:3}$, show very inconsistent patterns in detail. Long chains are generally absent, but when a linolenic acid is present it is the α rather than the γ isomer. More recent work by Yokokawa (1979) on six mushrooms in four genera confirms these general conclusions. The various species had principally $C_{16:1}$, $C_{18:1}$, and $C_{18:2}$. Similarly, among the Ascomycetes and Deuteromycetes chain lengths exceeding C_{18} are rare, and $C_{16:0}$, $C_{18:1}$, and $C_{18:2}$ are commonly the principal fatty acids. When $C_{18:3}$ is present, it is the $\Delta^{9,12,15}$ isomer (α-linolenic acid, the $\omega3$ isomer). The true yeasts tend to accumulate mono- (usually palmitoleate) rather than polyunsaturates, and the order Hypocreales, e.g., *Claviceps* sp., is characterized by yielding cases of hydroxylation and epoxidation of the unsaturates. The lowest fungi are set apart by frequently producing longer chain acids (C_{20} and C_{22}) and having the $\omega6$ isomer of polyunsaturates of either $C_{18:3}$ ($\Delta^{6,9,12}$ or γ-linolenic acid) as in the Phycomycetes or of $C_{20:2}$, $C_{20:3}$, and $C_{20:4}$ as in the slime molds. The lower Phycomycetes (classes Chytridomycetes and Oomycetes) also tend to have higher proportions of the polenoic acids with longer chains than do the higher Phycomycetes (class Zygomycetes). Curiously, the triunsaturate (γ-linolenic acid) of the lower fungi is a common acid in higher animals, illustrating that "higher" and "lower" can be an elusive concept. (See Wassef, 1977, for details of fungal fatty acids.)

The sterols of fungi show a variable pattern that is only in part associated with taxonomy (Nes and McKean, 1977). The most homogeneous group are the Ascomycetes and Homobasidiomycetes, in which ergosterol with varying amounts (which can exceed 50%) of its 5- and 22-dihydro and 5,22-tetrahydro derivatives is found. The methyl group of the sterol in both fungal groups has the 24β-orientation (Adler *et al.*, 1977). Ascomycetes and Basidiomycetes appear to be characterized by the presence of only the first 24-C_1-transferase in sterol biosynthesis, although suggestions that the second transferase (giving 24-ethylsterols) may be present in *Pullularia* and *Leptosphaeria* species exist (Weete and Laseter, 1974; Alais *et al.*, 1976). The spores of some rust fungi (Heterobasidiomycetes) definitely contain 24-ethyl- and 24-ethylidenesterols (see Nes and McKean, 1977), leading to a complicated problem in assigning clear genetic patterns. A still more complicated problem faces us with the other fungi. Dominant sterols can have no 24-alkylation, a 24-methyl group, or a 24-ethyl group. They can also have a Δ^5, $\Delta^{5,7}$, or Δ^7 type of unsaturation in ring B and a Δ^{22} or $\Delta^{24(28)}$ type in the side chain. No absolute taxonomic relationship exists, except that the Phycomycetes appear to have the greatest diversity. Thus, while most (but not all) species in the order Mucorales have a pattern (ergosterol and its 22-dihydro derivative) similar to that in the Homobasidiomycetes, in the orders Saprolegniales and Leptomitales 24-dehydro-, 24-methylene-, 24-methyl-, 24-ethylidene-, and 24-ethylcholesterol as well as 22-dehydrocholesterol, 22-dehydro-24-methylcholesterol, and 22-dehydro-24-ethylcholesterol have been isolated. The Deuteromycetes usually have the Homobasidiomycetous distribution, but not always. In this character, the otherwise quite different Deuteromycetes, Phycomycetous species in the order Mucorales, and Homobasidiomycetes are similar. The slime molds, which some class in a different phylum (Myxomycophyta) from the fungi (Eumycophyta), have, in the two genera studied (*Dictyostelium* and *Physarium*), no nuclear unsaturation. 24-H Sterols and 24-ethylsterols are found in *Dictyostelium*, and a mixture of 24-methyl- and 24-ethylsterols in *Physarium*. They are thus unusual and taxonomically consistent in having no unsaturation in the ring system but taxonomically inconsistent in the extent of alkylation at C-24. Further and especially notable are the fungi that seem to lack sterol biosynthesis completely and can grow vegetatively but cannot form oospores in the absence of sterol. Especially well studied are the pythiaceous fungi of the genera *Pythium* and *Phytophthora*. The absence of sterol biosynthesis may be a characteristic of the class Oomycetes and could well stand more extensive investigation both for phylogenetic and functional reasons. The primitive, parasitic *Plasmodiophora brassicae* also is reported to lack sterol biosynthesis. (For a deeper discussion and a key to the literature, see Nes and McKean, 1977, and Nes, 1977.)

It is perhaps worthy of note that fungi are not only quite diverse in their sterols and fatty acids but that there is no profound correlation between the two parameters of evolution, although certain relationships seem to emerge. The Homobasidiomycetes and Ascomycetes have a sterol pattern more or less restricted to 24β-methylsterols usually with a $\Delta^{5,7}$ grouping and also tend to restrict their fatty acids to the shorter lengths (not more than C_{18}) and to unsaturation not greater than two double bonds. The Phycomycetes (and to a lesser degree the Deuteromycetes), on the other hand, show greater diversity in their enzymic capabilities. Both the sterol and the fatty acid patterns have greater ranges in numbers of carbon atoms in the degrees of unsaturation. Such generalizations, however, may well be too crude to be useful. The fact that the spores of Heterobasidiomycetes have different sterols from the reproductive parts (mushrooms) of Homobasidiomycetes and that some fungi make no sterols at all counsels caution in our interpretation of the existing information.

The hydrocarbons of fungi have been recently reviewed by several investigators (Wassef, 1977; Weete, 1972; Rattray *et al.*, 1975). Unbranched and, to a lesser extent and variously with respect to species, branched chains are reported without clear taxonomic relationships. The chains vary as high as C_{39} with the odd numbers of C_{27}, C_{29}, and C_{31} quite common, but species such as of the genera *Penicillium* and *Aspergillus* are reported to have only even-numbered chains. *Trichoderma viride* is intermediate with both types of chain and an odd-to-even ratio of less than one. Individual alkanes are not as well identified as one would like. Polyunsaturated hydrocarbons on rare occasions, e.g., hexadecatriene in *Candida*, have been reported.

9. Lipids of Invertebrate Animals

Yamamoto *et al.* (1979) have examined the fatty acids of six species among the Gastropods (snails, phylum Molluska), Pelecypoda, and Crustacea (shellfish, phylum Arthropoda). The main fatty acids common to all species were n-$C_{14:0}$ and n-$C_{16:0}$, and the authors concluded that no obvious taxonomic differences were observable. More comprehensive investigations by Van Der Horst and his colleagues (Van Der Horst, 1970; Van Der Horst and Voogt, 1969; Oudejans *et al.*, 1971a,b; Van Der Horst *et al.*, 1972; Oudejans, 1972a,b; Van Der Horst and Oudejans, 1972; and references cited therein) have led Van Der Horst (1970) to the same general conclusion, but the basis of it was not so much the general similarities as it was the detailed differences. Snails (phylum Molluska), for instance, even among themselves do not show a homogeneous fatty acid pattern. Although the principal chains are even and unbranched, odd and branched chains are present, and the chain lengths (varying from C_{14} to C_{28}) and relative amounts of saturated to unsaturated chains are quite variable. Thus, in *Succinea putris* the main fatty acids are $C_{16:0}$ (18%), $C_{18:0}$ (17%), $C_{18:1}$ (18%), and $C_{18:2}$ (7%). In *Cepaea nemoralis* but not in *Arianta arbustorum* (both of which belong to the same family) there is a strong shift to longer chains. In *C. nemoralis* the fatty acid with the highest content was $C_{20:4}$ (17%), which was present only as traces if at all in *A. arbustorum* and *S. putris*. The amount of fatty acids in molluscs is low (1.5%), and this and other evidence, e.g., essential absence of glycerides, suggests they are present as membranous phospholipids. While they could in part arise from the diet, active biosynthesis from acetate occurs. *C. nemoralis* and *A. arbustorum*, incidentally, which are land snails, do not seem to accumulate squalene but do biosynthesize sterols, and *C. nemoralis* biosynthesizes unbranched and 2- and 3-methylalkanes (Van Der Horst and Oudejans, 1972). The distribution of alkanes (C_{15}–C_{37}) shows no preference for even or odd chains and is almost Gaussian with a peak centering at C_{27} in both the unbranched (major) and branched (minor) series. As discussed in Section D-5, the cuticular hydrocarbons of Insecta do not show such a continuous distribution. The meaning of the difference is unclear.

The African millipede, *Graphidostreptus tumuliporus*, in the class Myriapoda of the arthropods, constitutes an interesting example of the complications that arise in trying to use biochemistry as a chemotaxonomic and evolutionary marker with highly differentiated eukaryotic assemblages of cells. In the first place the *n*-hydrocarbons ranging from C_{15} to C_{36} of both males and females as well as of female eggs show, unlike the snails, a marked dominance of odd chain lengths (Oudejans, 1972a). The *n*-C_{29} compound stands out especially strongly above all the other components that include branched hydrocarbons. This suggests at least a partial dietary origin in higher plants and indicates that an analysis of the lipid content of animals for phylogenetic purposes must include the characteristics of gastrointestinal absorption and ecological factors as well as intrinsic biosynthetic capacity. The problem is not easy, though, since Oudejans (1972b) finds both sexes of this millipede will in fact biosynthesize hydrocarbons very actively from acetate as well as from branched precursors. Moreover, the fatty acid composition (C_{14}–C_{30}) of the two sexes is remarkably different in kind rather than just degree (Van Der Horst *et al.*, 1972; Oudejans *et al.*, 1971a,b). While both males and females have large amounts of *n*-$C_{16:0}$ (15–18%), *n*-$C_{18:1}$ (23–36%), and *n*-$C_{18:2}$ (8–12%) and also have small amounts of odd and branched acids, only females and their eggs contain cyclopropyl fatty acids (25% and 35%, respectively), which occur at the expense of mono-, di-, and triunsaturated acids. The acids with a three-membered ring have *cis* configurations and chains with a total number of C atoms of 17, 18, and 19. These cyclopropyl fatty acids appear to be structurally distinct from those found in bacteria, plants, and protozoa, which have only odd numbers of C atoms (Christie, 1970; Johnson *et al.*, 1967), and their manner of occurrence is different. In bacteria and protozoa they seem to be restricted to the phospholipid fraction, but in *G. tumuliporus* they occur as both phospholipids and neutral lipids. The structure, etc., of the millipede cyclopropyl acids would seem at first sight to constitute an excellent phylogenetic marker were it not that in another millipede, *Julus scandinavius*, cyclopropyl acids could be detected in neither males nor females (Van Der Horst *et al.*, 1972).

The fatty acids of larval and adult stages of the Insecta usually have straight chains with unsaturates in greater quantity than saturates, according to Svoboda *et al.* (1960) and the references cited therein. The $C_{16:0}$ and $C_{16:1}$ acids are the major components, for instance, in the housefly. The effect of taxonomy is small. The eggs of the grasshopper (*Aulocara elliotti*) are similar except for extension of the chain length and degree of unsaturation. The triglyceride fatty acids are mostly $C_{16:0}$, $C_{18:0}$, $C_{18:1}$, $C_{18:2}$, and $C_{18:3}$. The $C_{18:3}$ acid generally has the highest concentration (ca. 40%), but the relative proportions of the acids vary among different wild populations which in turn are different from animals reared experimentally. Cholesterol is usually esterified by an unsaturate that in the cockroach and housefly is oleate and in the grasshopper is linoleate. This tends to reflect the composition of the triglycerides with an emphasis on the principal unsaturate. Sterol fatty acid esters in plants also tend to reflect the other fatty acids. It is not yet clear what the membranous fatty acids of Insecta are.

Fast (1966) has made a monumental study of the phospholipids of 27 species of insect representing 20 families and six orders. Among the aphids (Homoptera) and in all but 1 family of the flies (Diptera), one-half of the lipid phosphorus was in phosphatidyl ethanolamine and one-quarter in phosphatidyl choline. In the one aberrant family of flies and in all the beetles (Coleoptera) examined, phosphatidyl ethanolamine and phosphatidyl choline were present in equal amounts. In all other insects phosphatidyl choline was dominant, with phosphatidyl ethanolamine composing 25–30% of the lipid phosphorous as is the case with most mammalian tissues studied. Of considerable interest is the fact that the fatty acid pattern in the phospholipids was shown to depend not on the species but on the esterifying group (choline or ethanolamine) on the phosphate moiety. As a first approximation the organismic occurrence of fatty acids with chain lengths less than 18 carbon atoms (especially palmitoleic acid) is associated with phosphatidyl choline, whereas the stearic acid is associated with phosphatidyl ethanolamine. Fast (1966) was led then to the important conclusion that the organismic distribution of fatty acids is not associated so much with taxonomy *per se* as it is with the type of phospholipid present and that this is related to specific, as yet not entirely clear in detail, functions of the phospholipid. The difference, of course, between choline and ethanolamine is a positive charge on the former. Since phospholipids act as membranous components, the functional association of the fatty acid component with the problem of charge will probably have to be sought in the membranes (cf. Section C-3).

10. Sterols of Eukaryotes

In the present section we will examine sterols more broadly than we have done in previous sections. Since the literature and detail of this subject is voluminous, the reader is referred to the earlier sections of this chapter and to Nes and McKean (1977) and Nes (1977) for more precise information. Before proceeding to eukaryotes we will summarize the situation in prokaryotes in order to place the subject in perspective.

Among prokaryotes, sterols are always found in blue-green algae but not always in bacteria. Perhaps of particular interest is the absence of sterols in photosynthetic bacteria, indicating that sterols are not a necessary component of photosynthesis. It is also interesting that in the nonphotosynthetic bacteria (*Methylococcus*, *Methylobacterium*, and *Azotobacter*) which have been most well investigated there is an accumulation of intermediates between squalene oxide and Δ^5-sterols. In the first two of the bacteria mentioned, sterols retaining one or more nuclear methyl groups are present, and in none of the three bacteria is a trisdesmethyl-Δ^5 sterol present, suggesting a primitive, in the sense of incomplete, development of the pathway. The blue-green algae, on the other hand, have sterols of the usual kind (4,4,14-trisdesmethyl-Δ^5 and -Δ^7 sterols), although there is a tendency not to complete the pathway in ring B. Thus, *Nostoc commune* has only Δ^7 sterols, *Calothrix* sp. has both Δ^5 and Δ^7 sterols, *Phormidium luridum* has Δ^5, $\Delta^{5,7}$, and Δ^7 sterols, and *Anabena cylindrica* has only Δ^5 sterols.

In eukaryotic algae there is a much greater tendency to complete the pathway to Δ^5 sterols. There are only a few species (e.g., in the genus *Chlamydomonas*, where one finds $\Delta^{5,7}$ sterols, and in the genus *Chlorella*, where there are species with Δ^7, those with $\Delta^{5,7}$, and those with Δ^5 sterols) that do not have primarily Δ^5 sterols. Also, unlike blue-green algae such as *Phormidium luridum*, the eukaryotic algae never have a strong mixture of Δ^7, $\Delta^{5,7}$, and Δ^5 sterols. That is, sterols of eukaryotic algae in a given cell line always have, except for very minor to trace components, only a given type of unsaturation in ring B. This tendency to complete the pathway (to Δ^5) as the complexity of the organism increases is continued in the tracheophytes, where only a few are known with primarily Δ^7 sterols, the family Cucurbitaceae being especially notable, and none is known to have $\Delta^{5,7}$ sterols as the dominant component.

Among nonphotosynthetic eukaryotes, some but not all have a completed Δ^5 pathway. Fungi with Δ^7 sterols include a great many Polyporaceae (Yokoyama *et al.*, 1975), the Basidiomycete, *Fomes applanatus*, and the Phycomycete, *Linderina pennispora*. Some have $\Delta^{5,7}$ sterols as in most of the Ascomycetes and Homobasidiomycetes, and some have Δ^5 sterols as in *Aspergillus oryzae* (Fujino and Ohnishi, 1979) and many Phycomycetes, especially of the orders Saprolegniales and Leptomitales. For a key to the literature, see Weete (1974). When we move to animals, we find the highest degree of all in completion of the pathway. Cholesterol is invariably the dominant sterol of the biosynthetic pathway in those animals in which a pathway exists, except for some of the Echinoderms (starfish, Asteroidea, and perhaps sea cucumbers, Holothuroidea) that make lathosterol (Δ^7).

The evidence, therefore, strongly suggests the following relationship. As the general complexity of the organism increases from a prokaryotic cell type to a eukaryotic one and then on to cellularly differentiated systems, the biosynthetic pathway develops both toward 4,4,14-trisdesmethyl-Δ^5 sterols as well as toward a kinetic control such that the end product of the pathway becomes the only sterol with a consequential steady-state concentration. An example of evolutionary ordering by these criteria would be to place the bacteria carrying the pathway only as far as squalene, e.g., *Halobacterium*, below those (*Methylococcus* and *Methylobacterium*) which cyclize squalene to lanosterol but incompletely demethylate the cyclized product. These in turn would be below *Azotobacter*, which seems to carry the pathway all the way to trisdesmethyl-Δ^7 sterols. The blue-green algae as a class must be moved to a higher level than the bacteria, because all blue-greens biosynthesize sterols and none accumulates sterols earlier in the pathway than the trisdesmethyl-Δ^7 stage. However, we can divide blue-greens into higher and lower groups. Those, e.g., *Nostoc*, that have only Δ^7 sterols would be more primitive than those, e.g., *Anacystis* and *Anabena*, that have only Δ^5 sterols. Species such as one studied in the genus *Phormidium* that have an array of Δ^7, $\Delta^{5,7}$, and Δ^5 sterols would have to be placed in between the other two. The eukaryotic algae as a class then can be placed above all the prokaryotes in that statistically one finds not only cyclization of squalene to sterols but completion of the pathway all the way to Δ^5 sterols to be much more frequent. Furthermore, rarely if ever is there the accumulation of intermedites. Overlap exists, though, with the blue-greens, in that some eukaryotic algae are Δ^7 producers and a few make only $\Delta^{5,7}$ sterols. Very little difference exists by this parameter (completion of the Δ^5 pathway) as we move from algae to higher plants—except in the detail of the statistics. Only in a small percentage of the cases examined, seemingly smaller than with the algae, are Δ^7 sterols the end product of the pathway, and in no case is a $\Delta^{5,7}$ sterol the end product. As with the eukaryotic algae, no tracheophyte strongly accumulates intermediates. All terrestrial photosynthetic plants from mosses to angiosperms (as well as those few angiosperms that are not photosynthetic) contain sterols, and in the many cases among the tracheophytes where incubations have been carried out the sterols are derived by biosynthesis rather than through absorption through the roots. Biosynthesis has been demonstrated in various plant parts, e.g., seeds and leaves, and is fundamentally independent of whether or not chloroplasts are present. The localization of sterols is also independent of ontogeny and the function of parts, since sterols are present in roots, stems, leaves, embryos, seedlings, mature plants, etc. The structures, therefore, of sterols in tracheophytes and probably also in the lower terrestrial embryophytes are indicative of what biosynthetic enzymes are present. This also seems to be true for the most part with the algae, but aquatic organisms always present the opportunity for absorption of sterols from the medium, which complicates the analysis of enzymic function. The freshwater algae examined have nearly always been grown experimentally, but the problem is acute with marine algae, where most of our knowledge comes from extraction of seaweeds that had grown naturally in the ocean. An important example is found with the red algae (Rhodophyta). These seaweeds always contain primarily or exclusively cholesterol, 5α-cholestanol, or more rarely the Δ^{22} or Δ^{24} derivatives of these two, yet many also contain sterols with shortened (C_7 instead of C_8 or more) or alkylated (C_9 or C_{10}) side chains (Chardon-Loriaux *et al.*, 1976). *Gelidium amansii*, for instance, has an array of 12 sterols (cholestanol being the major one) in the Δ^5 and Δ^0 series of nuclear unsaturation with side chains of C_7, C_8, C_9, and C_{10} some of which have Δ^{22} or $\Delta^{24(28)}$ unsaturation. Although the common dominance of cholesterol or cholestanol strongly indicates an enzymic pattern, it is by no means clear whether the other sterols are biosynthesized *de novo*.

     In the nonphotosynthetic line much the same thing is observed, although it is a good deal more complicated. In the first place stanols are sometimes present, and they can be the exclusive sterols in the organism as in the case of the slime mold *Physarum*. While stanols are known (rarely) in photosynthetic systems as minor components, they have never been found to represent the dominant sterol. Secondly, in the nonphotosynthetic line some of the organisms have the ability to assimilate sterols from the environment. This occurs in nonphotosynthetic organisms, especially those in the marine environment such as red algae as mentioned above, but only in nonphotosynthetic organisms does the assimilation of exogenous sterol reach proportions in which most of the sterol present is derived from another source. Total dependence on exogenous sterols is far more common than formerly appreciated. It occurs, among the fungi, Porifera, Protozoa, Coelenterata, Platyhelminthes, Nemathelminthes, Annelida, and Arthropoda with degrees of frequency varying from rare in the fungi, well documented only in the oomycetous genera *Phytophthora* ("plant destroyer") and *Pythium,* to essentially ubiquitous in the arthropods. Only in the Mollusca, Echinodermata, and Vertebrata is sterol frequently derived endogenously by *de novo* biosynthesis, and only in one of these, the Vertebrata, has a mechanism evolved for more or less complete gastrointestinal discrimination against exogenous sterol. Even in the Vertebrata the discrimination by gastrointestinal and other means has statistical variation, which has come to be of considerable importance in the medical management of human hypercholesterolemia in view of its statistical association with cardiovascular disease. As a result of the assimilation of exogenous sterols, we have to ask two questions in order to look at the problem of phylogenetics.

The first question has to do with what sterols are made in the cases where biosynthesis occurs. The answer is well documented. In vertebrates from fish to man the pathway is complete to cholesterol. Similarly, in the few cases investigated, molluscs biosynthesize Δ^5 sterols. The echinoderms are of two types. The Asteroides (starfish), and perhaps the Holothuroidea (sea cucumbers), biosynthesize lathosterol (Δ^7-cholestenol), while the Echinoidea (sea urchins), and perhaps the Crinoidea (sea lilies) and Ophiuroidea (brittle stars), biosynthesize cholesterol. One is tempted to place Δ^7 producers lower than those making Δ^5 sterols, but this may be a bit rash, because the starfish will convert exogenous Δ^5 sterols to Δ^7 sterols (as well as to stanols to some extent). This suggests that the stopping of the pathway at the Δ^7 stage is not primitive but purposeful, so much so that enzymes (Δ^5 to Δ^7) are present to guard against deposition of cholesterol, which would otherwise be regarded as the better material. This should then give us caution in making an absolute assessment of Δ^5 sterols as the highest type of sterol. Statistically it certainly seems to be, yet some organisms would prefer not to have it. Detail of function mated with ecology probably overrides a purely molecular or biosynthetic analysis. The failure of some organisms to prize Δ^5 sterols highly is found outside of the starfish. The protozoa, well examined in the genera *Tetrahymena* and *Paramecium,* dehydrogenate Δ^5 sterols to their $\Delta^{5,7}$ derivatives. In the former case only the $\Delta^{5,7}$ derivative is used for membranes, and in the latter case growth requires not only the $\Delta^{5,7}$ grouping but a 24-ethyl group. Cholesterol also will not support growth of some, though rare, insects, and among slime molds, it will be recalled, the pathway either bypasses Δ^5 sterols or goes beyond the Δ^5 to the Δ^0 stage.

The second question that has to be considered in an evolutionary ordering is the one associated with discrimination against exogenous sterols. The organisms below vertebrates in the classic scale of animal evolution have the least discrimination in sterol absorption. Some examples are as follows. The sponge, *Cliona celata,* has cholesterol together with as many as 8 other sterols mostly of the 24-alkyl type but also including a C_{26} sterol. A similar pattern is found in Coelenterata. The marine annelid *Pseudopotamilla occelata* has a mixture of 10 sterols comprising in decreasing order cholesterol (50%), 24-dehydrocholesterol (desmosterol, 18%), 22-dehydrocholesterol (6%), a series of 24-alkyl-Δ^5 sterols (20%), and two C_{26} sterols (5%). The termite, *Nasutitermes rippertii,* similarly has only 56% cholesterol, the remainder being sterols with shortened and (primarily) lengthened (24-alkyl) side chains. From the mollusc, *Placopecten magellanicus,* 17 sterols have been isolated, and no less than 22 sterols have been found in the starfish, *Asterias rubens.* The primitive chordate, *Halocynthia roretzi,* also has a complex mixture, but abruptly when the vertebrate line is encountered from fish to man, regardless of the diet (except in genetic abnormalities), the cholesterol content rises to a very high level (about 98%). From what is said here one would expect flatworms and flukes (Platyhelminthes) and roundworms (Nemathelminthes) to have sterol mixtures. They are not as well investigated as are many other forms of life, but, for instance, the nematode, *Ascaris lumbricoides,* is known to contain an array of Δ^0 and Δ^5 sterols in the 24-H, 24-CH_3, and 24-C_2H_5 series with cholesterol dominant. It has been possible with some of these organisms, notably Arthropoda and Protozoa but to a lesser extent also with Platyhelminthes and Nemathelminthes, to demonstrate absorption of sterols from the diet in contrast to extensive studies that show man poorly absorbs sterols from the diet especially if the sterol is not cholesterol. The ability to discriminate against sterols of exogenous origin seems from the above discussion to be a phylogenetic marker and is worth examining further.

There is a very strong statistical association between the animal line and the dominance of cholesterol in the animal, whether the source is endogenous or exogenous, and there is also a gradation in the quantitative amount of cholesterol in the sterol mixture as one passes from forms of life with very simple nervous systems to those with complicated ones. Thus, sponges contain at best a primitive nervous system and, while cholesterol along with other sterols is identifiable in some, e.g., *Cliona celata,* in others it is not present at all, as in *Hymeniacidon perleve,* which contains cholestanol along with many other sterols. *Axinella cannabina* even contains mostly sterols lacking the angular methyl group between rings A and B (19-norsterols in the Δ^0 and Δ^{22} series) together with *cis*-22-dehydrocholestanol. Equally unusual sterols (carbon added to C-26 or C-27, 5,8-peroxides, and five-membered ring A) have been found in other sponges. In the Coelenterata, where the colonial-cell type of organization of the sponges gives way to well-defined tissues and a clear nervous system, cholesterol is not only nearly always present but is usually though not always the major sterol. Moreover, while unusual sterols (alkylated at C-22 and C-23) are found in the class Anthozoa along with cholesterol, the occurrence of bizarre structures in the phylum Coelenterata as a whole seems to be less frequent and less to the exclusion of cholesterol than in the lower phylum Porifera (sponges). Not enough is yet known, though, to make too much out of the structure and statistics of occurrence of peculiar sterols except to say that in the future such sterols may become good phylogenetic markers and that they are found among many of the great taxonomic groupings of marine invertebrates but not in terrestrial life.

It would be satisfying to associate discrimination against exogenous sterols with evolution of body plans. A good candidate would be the development of a coelom. Unfortunately, eucoelomate animals such as the molluscs have as large an array of exogenous sterols as is known in any organism, while eucoelomate animals also include man, who has an extremely small capacity to assimilate exogenous sterol, especially if it is not cholesterol. On the other hand, two body plans do stand out. One is the vertebrate line, in which biosynthesis occurs and leads to cholesterol and in which there is also strong discrimination against exogenous sterol. In addition, plant sterols that succeed in invading vertebrates are metabolized more rapidly than cholesterol, presumably as a further protection to prevent any sterol other than cholesterol from being present. The other body plan, already alluded to, is the presence of a nervous system itself, the very criterion by which we would define an animal. Statistically, Δ^5 sterols, more precisely cholesterol itself, are found in animals rather than something else. Thus, the vertebrate line has a unique homogeneity in both completing the Δ^5 pathway and discriminating against exogenous sterol. Since other animals may have one of these characters but not both, the vertebrate line clearly is a special one and higher (in the sense of having evolved more extensive regulation of what sterol is present) than other creatures with nervous systems. While lower animals are less easy to order in this way with respect either to each other or to the fungi and slime molds, again the presence of Δ^5 sterols, especially the dominance of Δ^5 sterols, on a statistical basis is strongly favored in the animal line. In the molluscs, for instance, despite the presence of many different sterols by ingestion and a large range for the amount of cholesterol (19-95%) in the Gastropods, Bivalvia, and Cephalopoda, the average value (65%) for the amount of cholesterol in the sterol mixture definitely favors cholesterol. Furthermore, except for unusual cases already discussed, true animals either make or prefer Δ^5 sterols. This contrasts sharply with nonphotosynthetic systems lacking a nervous system, such as those yeasts, slime molds, and protozoa that neither make nor prefer Δ^5 sterols. The data, in summary, for the eukaryotic nonphotosynthetic line do roughly parallel what is known in the photosynthetic line, namely, that the more highly differentiated an organism is, i.e., the more complex it is, the more likely we are to find the occurrence of Δ^5 sterols. We will return to this again in the section on biosynthesis. For the present we should like to leave the subject with the reminder that even though the biosynthetic data are much more precise and extensive there seems to be a preference for Δ^5 sterols even in the absence of biosynthesis as the evolutionary ladder is climbed.

The relationship between a preference for a given type of exogenous sterol and the animal line and the correlation of this with the biosynthetic pattern is much better documented in terms of the sterol side chain than it is with respect to the structure of the nucleus. Animals are very clearly differentiated from nonanimals by a nearly absolute dependence on sterols lacking an alkyl group at C-24. Animals throughout the whole evolutionary range have been studied, and no animal tissue has been found that will introduce an alkyl group. The side chain representing the terminal eight carbon atoms of squalene which have not been cyclized and have only been reduced or dehydrogenated is unquestionably not only what animals make but what they require. The requirement has been documented in several different systems. Man selectively absorbs cholesterol from a mixture of cholesterol and 24-alkylcholesterol. Rats absorb 24-methylcholesterol better than 24-ethylcholesterol, and White Carneaw Pigeons exhibit the same selectivity. Similarly, the housefly (*Musca domestica*) absorbs cholesterol better than 24-methylcholesterol, which in turn is absorbed better than 24-ethylcholesterol. Clearly the mechanism of absorption is tuned to 24-desalkylsterols. This is all the more interesting when it is remembered that insects cannot biosynthesize sterols. The absorption is consequently not tuned to biosynthesis. The only other reasonable assumption is that both it and biosynthesis are tuned to function. Accepting this we then get a correlation with the general occurrence data. Forms of life having a nervous system tend to have a sterol lacking a 24-alkyl group. This functional requirement, best met in the Δ^5 series as cholesterol, is encountered to its fullest extent abruptly in the lowest of the vertebrates (fish) well studied. This conclusion follows from the fact that discrimination against other sterols is not high in *Halocynthia roretzi*, even though it has cholesterol. The latter organism (a tunicate) is a primitive chordate but not a vertebrate, and its sterols include only 43% cholesterol and other 24-desalkylsterols (desmosterol, lathosterol, and cholestanol). The remainder is composed principally of sterols bearing extra C atoms at C-24 (mainly 24-methylenecholesterol) and to a lesser extent (7%) by C_{26}-sterols. Animals below the chordate level of development also commonly have cholesterol in a complex sterol mixture that, as in the tunicate, usually includes a large amount of 24-alkylsterol. In the scallop, for example, *Placopecten magellanicus,* of more than a dozen sterols found cholesterol had just barely the highest amount (less than 50% of the total), and the next most abundant sterol was 24-methyl-22-dehydrocholesterol, followed by 24-methylenecholesterol. The various sterols with 24-C_2 or 24-C_3 groups were each present in much smaller amounts than those with smaller side chains. This situation parallels absorption data for vertebrates in terms of the relative size of the side chain, but the absolute extent of the discrimination is less in the lower animals. In view of the importance of cholesterol the animals that lack sterol biosynthesis and feed on sources that have *only* 24-alkylsterols, as with some plants, should, so to speak, be in some trouble. Actually, they are not. How they handle this matter is one of the most compelling reasons to believe that the lack of alkylation at C-24 is functionally associated with a nervous system. What the majority of insects ingesting only 24-alkylsterols (many herbivorous species) do to circumvent this problem is to have the enzymatic apparatus to dealkylate these materials, yielding cholesterol.

Photosynthetic systems are divisible morphologically into the lower ones (algae), which do not reproduce with the intermediacy of a well-defined embryo, and the higher ones (embryophytes), which do. The embryophytes can then be shown to be subdividable into groups that do not (Bryophyta) and those that do (Tracheophyta) have true roots and a true vascular system. The latter are further classifiable by increasing sophistication in reproduction, as in the sequence from sporeformer, such as the ferns, to the multitudinous pteridophytes, which in depositing food along with the zygote give a form of life we call seeds. The seed-bearing plants then are in two major, ascending categories, viz., those in which the seed is poorly protected (gymnosperms or "naked seeds") and those that build structures around the seed (angiosperms). The latter also have a complex organ, the flower, in which the male and female parts of the plant are situated, sometimes alone and sometimes together. The appearance of these various types of plant in the fossil record roughly parallels their degree of complexity. More precisely, the coal-forming swamps in the Carboniferous left abundant fossils of the low tracheophytes, e.g., ferns and lycopods, while only later are fossils of angiosperms well documented. While the fossil record leaves much to be desired, the approximate correlation between time of appearance and morphological complexity has classically suggested an evolution of a common gene pool. That is, it is attractive to suppose that algae led to mosses, which led to ferns, which led to gymnosperms, which led to angiosperms. Some people are fond of stating this sequence with a caveat by saying that "something like" a moss led to "something like" a fern, etc. In this way one avoids implying that a particular species, especially one now extant, was the direct progenitor of some other. Fundamentally, however, this does not alter the supposition that embryophytic evolution began in algae and passed from a nonvascular to a vascular state and that current representatives of these evolutionary stages exist. These ideas are supported by the sterol patterns (Nes *et al.*, 1975b, 1976b, 1977a; Catalano *et al.*, 1976; Nes, 1977; Nes and McKean, 1977), but only if we take a statistical view of the subject, and even under these circumstances a number of problems are left unresolved.

Starting with the mosses, we find several interesting things. In the first place, intermediates bearing nuclear methyl groups frequently accumulate. Among these are cycloeucalenol, obtusifoliol, and 31-norcyclolaudenol, all of which have a 4- and a 14-methyl group and a methyl or methylene group at C-24. The amounts of these three intermediates alone reach 78% of the sterol mixture in *Racomitrium lanuginosum* and 38% in *Campylopus introflexus.* Although in *Ctenidium molluscum* 97% of the sterols lack nuclear methyl groups, four of the five species studied, each representing a separate genus, had substantial amounts of intermediates with nuclear methyl groups. Moreover, with three of the five, including *Ct. molluscum,* 20–30% of the sterol possessed the $\Delta^{5,7}$ grouping in the nucleus and a Δ^{22}-24-methyl arrangement of the side chain (ergosterol or its 24-epimer, the configuration having not been elucidated). The other sterols present that lacked nuclear methyl groups were principally in the Δ^5-24-ethyl series with smaller amounts of Δ^5-24 methylsterols. The failure to complete the Δ^5 pathway in a kinetic sense is extremely unusual, being completely unknown in algae, and very uncommon in angiosperms. On the other hand, it reminds us of the bacteria. There is the remarkable suggestion here that the evolutionary process has started all over again. Such an idea is given further credence by the fact that the low tracheophyte, *Lycopodium complanatum,* also contains a substantial amount of $\Delta^{5,7}$ sterol. The 4-desmethylsterols in decreasing order from the plant collected in the summer were ergosterol, 24α-ethylcholesterol, 22-dehydro-24α-ethylcholesterol, 24β-methylcholesterol, and perhaps small amounts of the 24α-methyl compounds, 24-epiergosterol and 24α-methylcholesterol. The ergosterol content was reduced in a fall sample. Configurational assignments are quite secure in the lycopod case.

It will be seen that lycopod and mosses have closely similar sterol patterns. Assuming that sterols in mosses have lycopod configurations, we would find *Ct. molluscum,* for instance, to have 24% ergosterol, 54% 24α-ethylcholesterol, and 19% 22-dehydro-24α-ethylcholesterol. The investigated ferns are primarily Δ^5 producers, and have no $\Delta^{5,7}$ sterols, but in one (*Dryopteris noveboracensis*) of four cases significant amounts of Δ^7 sterols (lathosterol and its 24α-ethyl derivative) have been found. In the presence of Δ^7 sterols however, this frequency approximates that in higher plants. Thus, the ferns seem to be evolutionarily higher than the lycopods.

As the ladder is climbed to gymnosperms and angiosperms, the majority of many plants carefully examined in a variety of genera and orders prove to have a uniform sterol pattern of the following sort. The principal component is 24α-ethylcholesterol (sitosterol), on occasion accompanied by its 22-*trans*-dehydro derivative (stigmasterol). These two 24α-ethylsterols are never accompanied by their 24β-epimers. The next most abundant constituent is 24-methylcholesterol, followed by cholesterol itself. The 24-methyl component, unlike its 24-ethyl analog, is always an epimeric mixture of 24α-methyl-(campesterol) and 24β-methylcholesterol (dihydrobrassicasterol) in a ratio of about 2:1, respectively. In some cases, e.g., certain *Euphorbia* and *Solanum* species, the amount of cholesterol is very high, while in many others it is only a trace component. Plants with these sorts of sterol mixtures can be generalized as having the homologous series of cholesterol and its 24-methyl and 24-ethyl derivatives with the 24α-alkyl configuration dominant when C-24 bears a substituent. These have been designated as plants of Category I-A (I for 24α dominance and A for Δ^5) and called "main line" plants (Nes *et al.*, 1977a). Category I-B is the designation for plants primarily producing 24α-Δ^7 sterols, and Categories II-A and II-B are given to plants producing primarily 24β-alkylsterols in the Δ^5 and Δ^7 series, respectively. The rarer categories I-C or II-C are reserved for plants with principally $\Delta^{5,7}$ sterols. Although some of the mosses and lycopods approach being in Category C, a true C condition is only known in the algae and fungi.

As already mentioned, most of the investigated ferns, gymnosperms, and angiosperms are main-line plants. This suggests that the introduction of a 24α-alkyl group represents maximal evolution up to the present time, and that selection for this type of sterol is greater the larger is the group at C-24. Sitosterol with a Δ^5-24α-ethyl structure would be the most highly evolved compound. Such a view would not mean that the 24β-alkylsterols do not function, because they unquestionably do. It also would not mean that they are not mated to other molecules in a lock-and-key fit. The idea does suggest, however, that plants having only sitosterol and the appropriate receptors may represent a more highly evolved condition than the others which exist. Some plants, e.g., cotton based on its seed oil, which is 93% sitosterol, approximate this elevated state, but none seems to reach it entirely. A more exacting analysis of the detailed role of various sterols is obviously needed before too much credence can be placed in this suggestion; yet the triple parameter of size of substituent, configuration of substituent, and nature of unsaturation in ring B does have some evolutionary correlates—as well as problems.

Again taking a statistical approach, we find that the frequency of occurrence of 24β-alkylsterols seems to be very much greater in algea than in embryophytes. There is, in fact, only one algal type (the diatoms) in which 24α-alkylsterols are known to occur. All others studied are 24β producers exclusively. Conversely, there are only two known tracheophyte families (Crassulaceae and Verbenaceae) in which 24β-alkylsterols occur exclusively. Satisfyingly, lycopods, which are intermediate between algae (primarily

Category II) and the gymnosperms and angiosperms (primarily Category I), are a mixture of Categories I and II. If the ergosterol in mosses is really ergosterol as it is in lycopods, then the mosses also are intermediate as one might hope. There thus appears to be a nice correlation: the more primitive an organism, the more likely it is to be of Category II. This leads to the prediction that the 24-alkylsterols of blue-green algae will be found to have the 24β configuration. It is not clear what one can do when the statistical approach is not used. In the simplest sense, as the evolutionary ladder is climbed we would like to have seen the regular and logical development of Δ^7, $\Delta^{5,7}$, and Δ^5 sterols i.e., plants should proceed from Category B to C to A, since this is the way the sterols are formed in the pathway. Although there is no way *a priori* of deducing which configuration (α or β) at C-24 should be first, the empirical data indicate the β configuration is more primitive. Consequently, we might anticipate that plants of Category II would always be lower than those of Category I. This, however, is clearly not always the case, or more exactly it is not the case when we use classic, mostly morphological, parameters of evolution as discussed in what follows (Nes *et al.*, 1977a).

Two quite different taxonomic and phylogenetic relationships have been suggested in the literature for the angiosperms by Hutchinson (1969) and Cronquist (1968). Chemotaxonomy based on the sterols offers a way of examining them. From morphologic, geographic, and other parameters, both Hutchinson and Cronquist place Magnoliales (including *Liriodendron tulipifera*) as the most primitive order of flowering plant closely associated with Ranales of the Hutchinson system, which Cronquist (at least in so far as *Podophyllum peltatum,* which is in Category I-A, is concerned) regards as Ranunculales. Hutchinson considers Magnoliales to have given rise to woody dicots and Ranales to herbaceous dicots and perhaps also to monocots, while in the Cronquist system the woody and herbaceous divisions are not recognized. Since the family Cucurbitaceae and the examined species of Theales are of Category B and both Magnoliales and Ranales (Ranunculales) are of Category A, the sterol data correlate with neither system in the sense of a direct progenitor role of Magnoliales for these Category B plants. Nes *et al.* (1977a) suggest that the sterols may indicate a parallel evolution with, rather than a direct evolution from, Magnoliales. Similarly, Hutchinson's suggestion of the evolution of Saxifragales from Ranales is not verified, since *Kalanchoe daigremontiana* (Saxifragales) is of Category II and *Podophyllum peltatum* (Ranales) is of Category I; nor is his suggestion of the line to Ericales through Theales from Magnoliales verified, since *Kalmia latifolia* (Ericales) is of Category I-A, while representatives of Theales, e.g., the genus *Camellia*, are of Category I-B. The Theales to Ericales (and Capparales including *Brassica*) line of Cronquist similarly is not consonant with the data. On the other hand, there are aspects of the Cronquist system with which the data either do agree or with a slight change in his system would agree. He places *Cucurbita* in Violales emanating from Theales, and both are, in fact, of Category I-B in the mature stage. He also places Crassulaceae (which includes *Kalanchoe*) in the Rosales with Leguminosae (which includes *Pisum* and *Glycine,* both I-A), and the former is supposed to have given rise to the latter, with which the change from Category II (*Kalanchoe*) to Category I (*Pisum* and *Glycine*) would agree. *Clerodendrum* (Category II-A) in Lamiales is placed above Rosales. Again the sterol data agree, but only if the evolutionary line to *Clerodendrum* (Category II-A) bifurcates (at or before Crassulaceae, which is of Category II-A) before the formation of Leguminosae which is Category I-A. Unfortunately, despite the coincidence of the data in and above Rosales with the previous classifications, the sterol data do not support the origin of Rosales itself (in the Cronquist system) from Magnoliales (Category I-A), in view of his placing Crassulaceae (Category II-A) in the order. In Hutchinson's system, Crassulaceae is supposed to have emanated from Ranales, which agrees no better.

The detailed disagreements delineated between sterol structure and previously suggested phylogenetics should not, however, obscure more general agreements. The sterol structures are consistent with, while not proving, the idea that many angiosperms are higher and could have arisen from the gymnosperm, since the two species of the latter studied are of Category I-A as are most of the angiosperms studies. Similarly, the Filicopsida are of Category I-A. This tells us that evolution of the pathway to 24α-ethylsterols in the Δ^5 series was accomplished quite early chronologically and is not associated with the presence of characters such as flowering. In summary, if retroevolution did not occur in the sterol pathway, which at the present is not amenable to experimental verification, some of the angiosperms, e.g., *Cucurbitaceae,* do not seem to have arisen from a line comprising Filicopsida, Gymnospermae, and Magnoliales and may have diverged prior to the principal line, but the sterol data do not conflict with the thesis that many flowering plants could have arisen from Magnoliales. An alternative view, of course, is that function rather than phylogenetics has determined what type of sterol is biosynthesized.

CHAPTER 8

Phylogenetics and Biosynthesis

A. STEROL BIOSYNTHESIS

1. Photosynthetic vs. Nonphotosynthetic Routes

The sterols of living systems are all derived by the aerobic cyclization of squalene through the intermediacy of epoxysqualene as discussed in depth elsewhere (Nes, 1977; Nes and McKean, 1977). However, epoxysqualene is not always cyclized to the same substance. Two major products (lanosterol and cycloartenol) are formed, and they are never known to be produced by cyclization in the same organism, whether the latter is unicellular or highly differentiated. This bifurcation in the steroid pathway is one of the most interesting phylogenetic markers we have, because it has no apparent influence on the structure of the functional steroid at the end of the pathway. There are two principal differences in naturally occurring, functional sterols, viz., the structure of the side chain and the double-bond character in the nucleus. Neither of these is influenced by whether lanosterol or cycloartenol is the precursor. Thus, it was shown in our laboratory (Russell *et al.*, 1967; Raab *et al.*, 1968; Gibbons *et al.*, 1971) that either lanosterol or cycloartenol will yield 24-alkylsterols (sitosterol, etc.) in higher plants. Similarly, while lanosterol leads to cholesterol in animals (Tchen and Bloch, 1955), pollinastanol (14α-methyl-9,19-cyclo-5α-cholestanol, a metabolite of cycloartenol in which the Δ^{24} bond has been reduced and the two methyl groups at C-24 removed) has been converted to cholesterol in higher plants (Devys *et al.*, 1969), and cycloartenol is present in red algae in which the dominant sterol is cholesterol (Ferezou *et al.*, 1974). On the other hand, lanosterol leads to $\Delta^{5,7}$-24-alkylsterols in fungi (Schwenk and Alexander, 1958). This disassociation

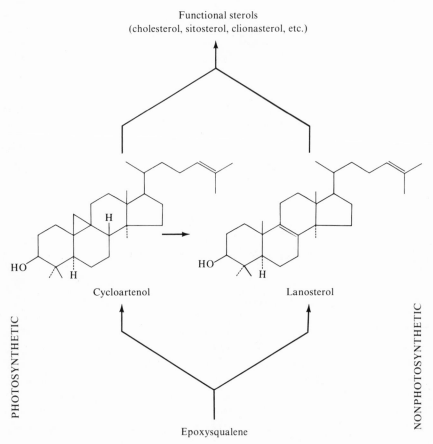

Figure 21. The cycloartenol–lanosterol bifurcation.

of the structure of the intermediate from the structure of the functional sterol renders the cycloartenol–lanosterol bifurcation independent of ecological and related pressures. These pressures would have to impinge on the functional products (cholesterol, sitosterol, ergosterol, etc.) directly or on factors altering their structures. Put in another way, evolution by interaction with and adaptation to an ecological niche should have no correlation with the cycloartenol–lanosterol bifurcation, since the latter has no bearing on what sterol is ultimately formed. Consequently, the bifurcation should reflect phylogenetic lines independently of the ultimate nature and character of the organism in question. The bifurcation is shown in Figure 21. The sequence of events following formation of lanosterol or cycloartenol depends on the organism involved. Cycloartenol differs from lanosterol in the presence of a three-membered ring (9,19-cyclo grouping) in place of a Δ^8 bond. This ring is opened in an energetically downhill process in the presence of an isomerase that protonates C-19 and deprotonates C-8 yielding a Δ^8 sterol (Rahier *et al.*, 1977). Thus, cycloartenol yields lanosterol, and cycloeucalenol leads to obtusifoliol.

It is possible *a priori* that there could be a random distribution of the two kinds of cyclase (epoxysqualene–lanosterol–cyclase and epoxysqualene–cycloartenol–cyclase), but this is not the case. Instead, there is a clear association of cycloartenol with photosynthetic systems and lanosterol with ones lacking photosynthesis. This was first fully appreciated in our laboratory (Gibbons *et al.*, 1971) following the suggestion by Ourisson and his colleagues (Benveniste *et al.*, 1965, 1966) that cycloartenol probably replaces lanosterol in tobacco plants. A full account of the history and of the extensive evidence for the phylogenetic association is to be found elsewhere (Nes, 1977; Nes and McKean, 1977). Some crucial points, however, are as follows.

We were able to show that the cycloartenol route is not confined to chloroplastic tissue of a photosynthetic plant, since the endosperm of pine seeds, which lacks chloroplasts and is completely white throughout its life, still uses the cycloartenol rather than the lanosterol pathway (Malhotra and Nes, 1972). That the cycloartenol pathway operates in the absence of the lanosterol pathway in chloroplastic tissue has been demonstrated in various algae, e.g., *Ochromonas malhamensis*, (Rees *et al.*, 1969) and *Chlorella* (Doyle *et al.*, 1972). The evidence for operation of the cycloartenol pathway in differentiated photosynthetic plants both in terms of the whole plant or certain plant parts is especially extensive. In mosses, for instance, cycloeucalenol (14-desmethyl-24-methylenecycloartanol) is present as a marker of the pathway, and various kinds of evidence with many higher plants, including gymnosperms and angiosperms, show the cycloartenol pathway operates. Of particular interest are investigations showing the lanosterol pathway fails to operate in the presence of the cycloartenol pathway in a gymnosperm (Malhotra and Nes, 1972) and several angiosperms (Gibbons *et al.*, 1971; Rees *et al.*, 1968; Ehrhardt *et al.*, 1967; Eppenberger *et al.*, 1969). The converse is true in animals and fungi (Mercer and Johnson, 1969). Even in photosynthetic plants (Euphorbiaceae) when lanosterol can be detected, cycloartenol is present and is demonstrably metabolized to lanosterol (Ponsinet and Ourisson, 1968), indicating the cycloartenol pathway is in operation. Conversely, the lanosterol pathway operates in the absence of the cycloartenol pathway in fungi (Anding *et al.*, 1974), mammals (Gibbons *et al.*, 1971), and nonphotosynthetic, methane-oxidizing bacteria (Bird *et al.*, 1971). In the first two cases cycloartenol fails to proceed to the functional sterol (ergosterol and cholesterol, respectively) while lanosterol does, and in methane-oxidizing bacteria lanosterol or its derivatives but not 9,19-cyclo compounds are present.

The evidence cited compels us to believe at least tentatively that there are two great evolutionary lines, a photosynthetic family and one lacking photosynthesis, which transcend both the degree of cellular differentiation and the presence or absence of a nervous system. Methane-oxidizing bacteria, fungi, and animals are in one family and algae, mosses, and tracheophytes in another. The true phylogenetic rather than functional character of the bifurcation has been demonstrated in a way that simultaneously traces phylogenetics in otherwise equivocal cases. Euglenoids can operate either autotrophically in a green form or heterotrophically in a leuco form. Now, if the bifurcation were associated with function, then the green form should operate the pathway via cycloartenol, whereas the lanosterol route should proceed in the leuco form. Actually, in both forms of *Euglena gracilis* the cycloartenol pathway operates (Anding *et al.*, 1971) as it also does in the nonphotosynthetic euglenoid *Astasia longa* (Rohmer and Brandt, 1973). It follows that *Euglena* and *Astasia* are phylogenetically photosynthetic. The leuco form is a special case of the green form and not the other way around. In higher plants the same conclusion can be arrived at. There are a few completely white angiosperms with no chloroplasts and no photosynthesis. Perhaps the most well known are "Indian pipes." Among the others, several have been investigated, viz., *Cuscuta europaea* (dodder), *C. epiphymum*, *Orobanche lutea* (broomrape) (Rohmer *et al.*, 1975), *Monotropa hypotitys* (Stanley and Patterson, 1977), and *Epifagus virginiana* (Nes *et al.*, 1979a), and all utilize the cycloartenol route. While not functionally photosynthetic, these plants clearly belong to the photosynthetic kingdom phylogenetically. Similarly, *Saprolegnia serax*, an Oomycete with a motile phase belonging to the Mastigomycotina in the taxonomy of fungi, converts cycloartenol to cholesterol, fucosterol, desmosterol, and 24-methylenechlesterol, indicating the presence of a 9(19)-cyclo-to-Δ^8 isomerase and a phylogenetic relationship to photosynthetic plants (Bu'Lock and Osagie, 1976). Other markers of a nonfungal origin of this organism are cellulose in the cell wall and synthesis of lysine by the higher plant pathway via diaminopimelic acid. Using the bifurcation, investigators have also thrown considerable doubt on the idea that higher plants are evolutionary symbionts of fungi and algae. If this were the case, only chloroplasts should utilize the cycloartenol pathway, but as mentioned earlier, a nonphotosynthetic part of a photosynthetic plant uses only the cycloartenol route (Malhotra and Nes, 1972).

The evidence in toto weighs heavily against not only the symbiont theory but also against random mutation. There is no apparent biochemical reason why a mutation from the cycloartenol to the lanosterol route could not have taken place. In the several cases studied, e.g., with *Epifagus virginiana* (Nes *et al.*, 1979a), *Pinus pinea* (Raab *et al.*, 1968), and the algae *Ochromonas malhamensis* (Lenton *et al.*, 1971) and *Chlorella ellipsoidea* (Tsai and Patterson, 1976), lanosterol will proceed in photosynthetic systems to functional sterols as well as will cycloartenol, yet none of the investigated photosynthetic systems utilizes the lanosterol route. Only a single enzyme (the cyclase) would have to be changed. The subsequent enzymes in the pathway would not have to be changed, since they are not specific for the 9,19-cyclo grouping. The converse is not true. Neither yeast nor rats will accept cycloartenol in place of lanosterol even for an initial step with the possible exception of reduction of the Δ^{24} bond (Gibbons *et al.*, 1971; Anding *et al.*, 1974). The enzymes in these otherwise different organisms are related both in the kind of cyclase and in the absence of an isomerase for opening of the 9,19-cyclo ring as well as in the binding constants of subsequent enzymes in the pathway. This further verifies a family relationship for most nonphotosynthetic organisms.

The possibility cannot be completely ignored that the bifurcation is actually extended to a trifurcation. The $\Delta^{9(11)}$ analog (parkeol) exists in some photosynthetic plants (Itoh *et al.*, 1975, and references cited), and in the alga *Ochromonas malhamensis* parkeol proceeds to functional sterols (Palmer *et al.*, 1978). Parkeol is theoretically an alternative product of cyclization of epoxysqualene in which elimination of a proton occurs from C-11 instead of C-19. However, our understanding of the mechanism of opening of the 9,19-cyclo grouping also permits parkeol to be a metabolite of cycloartenol (as an alternative to the conversion of cycloartenol to lanosterol). On balance, the more likely of the two origins of parkeol seems to be via cycloartenol, but the problem deserves additional attention. Euphoids, incidentally, definitely do not represent an alternative. We (Nes and Sekula, unpublished observations) have found the Δ^5 sterols of a species of *Euphorbia* (the whole poinsettia plant) to be the same stereochemically and otherwise as in other main-line plants except for an unusually high proportion of cholesterol. The euphorbs, while containing 4,4,14-trimethyl euphoids in the latex, clearly do not use these substances as intermediates to stereoisomers of the 4,4,14-trisdesmethyl-sterols. Cyclization of epoxysqualene to euphoids therefore seems to be functionally associated with latex, at least in the Euphorbiaceae, and does not seem to relate to a wider phylogenetic line.

The next subsection, which deals primarily with plants vs. animals, also includes a substantial consideration of evolutionary lines that depend in part on the ideas presented in relation to the cycloartenol–lanosterol bifurcation. For a deeper understanding of this bifurcation the interested reader should consult what follows.

2. Plant vs. Animal Routes

Both in science and in general human awareness the world was first viewed as two kingdoms, scientifically the kingdom Animalia and the kingdom Planta. The phylogenetic association that the cycloartenol–lanosterol bifurcation gives us (Section A-1) strongly indicates a still more fundamental division into the kingdom Photosynthetica and the kingdom Nonphotosynthetica, with animals falling into a subcategory of the latter and plants into both kingdoms. Let us examine this further. In particular, what constitutes being an animal or a plant? In one view the answer lies simply in the presence or absence of a nervous system. With a nervous system many avenues for existence are permitted that cannot be achieved without one. This is a functional matter, and the chemical requirements of a nervous system per se should lead to a certain degree of commonality among animals. In the simplest sense we might expect all nervous systems to have the same basic components. The remarkable fact from an evolutionary point of view is that this is not only true, but it is true in all tissues of animals, whether they are neural or nonneural tissues. We are specifically speaking here of cholesterol, which is biosynthesized not only by the mammalian brain but by all other mammalian tissues. The genetics for cholesterol are ingrained in the whole organism. Moreover, cholesterol is not only biosynthesized in many animals; it is required by the vast majority of those that do not carry out their own biosynthesis. Why is this remarkable? At a biochemical level, we know that cells can give other kinds of sterols. The most widespread process is to alkylate the Δ^{24} bond of the side chain that remains from squalene after cyclization. Alkylation is quite independent of whether an organism is photosynthetic or nonphotosynthetic. It proceeds in most photosynthetic organisms as well as in fungi, nonphotosynthetic algae, and nonphotosynthetic tracheophytes, but never, to our knowledge, in animals. Animals reduce the Δ^{24} bond, while plants either reduce it or alkylate it or carry out both processes. We have here a very precise molecular event by which an enormous category of living systems can be categorized. More exactly, it is a restriction in molecular events for organisms with a nervous system. Animals have open to them a single route, viz., to reduce the double bond (Δ^{24}), while plants have the opportunity not only to reduce but to alkylate it in several ways. This has been called the reduction-alkylation bifurcation (Nes, 1977; Nes and McKean, 1977). It is shown in Figure 22.

Before passing to the detail of this subject that permits among other things a phylogenetic analysis within the plant kingdom, we should like to stress that the restriction to reduction found in animals is not the only restriction known in creatures with a nervous system. The number of different kinds of lipids and related compounds in plants is generally far greater than in animals. Although it is also true with many biochemical classes, e.g., carbohydrates, in the isopentenoid class alone, the difference between plants and animals is enormous. There are tens of ways by which squalene and its oxide are cyclized in plants, giving a bewildering array of tetra- and pentacyclic products. Animals make only one, the sterols. At other levels of

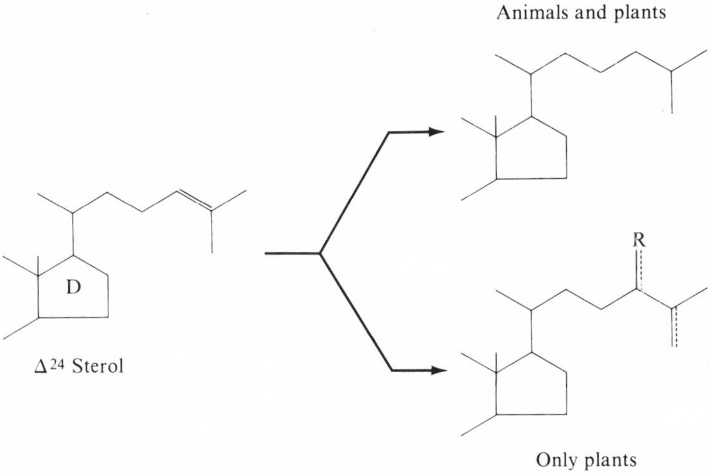

Animals and plants

Δ^{24} Sterol

Only plants

Figure 22. Reduction-alkylation bifurcation. Interrupted lines show where double bonds can be introduced following C_1 transfer to C-24 or in a second process to C-28.

polymerization of the C_5 unit, the same is true. Plants biosynthesize a vastly larger number of isopentenoids (terpenes, sesquiterpenes, sesterterpenes, diterpenes, etc.) than do animals, even though some animals (arthropods) make a few of the same compounds. The restriction is especially evident in the vertebrate line, where the isopentenoid pathway has but a very few courses. These are quantitatively dominated by biosynthesis of cholesterol, but dolichol, ubiquinone, and perhaps carotene in the corpus luteum are among the few other products of the vertebrate pathway. Also of phylogenetic importance is that the vertebrate line diverges from certain other animal lines in having always an intact system for the *de novo* biosynthesis of cholesterol. This ability is shared with some of the other animal phyla (Mollusca, Echinodermata, some Porifera, and marine Annelida) but not apparently with others (terrestrial Annelida, all or most of the Arthropoda, Nemathelminthes, Platyhelminthes, Coelenterata, and some Porifera).

If one takes the speculative sum of morphologic and fossil evidence (Valentine and Campbell, 1975), there is an interesting rough correlation with the biosynthetic information. It suggests that if there were a common origin of animals, which is by no means certain, sterol biosynthesis was acquired after the onset of the Cambrian. The Coelenterata, for instance, and the Platyhelminthes, both of which appear to lack the pathway, are thought to have diverged from the main stream some 200 million years prior to the Cambrian, i.e., 800 million years ago. The Porifera, Mollusca, and Annelida are believed to have arisen more or less in the Cambrian (600 million years ago), and most but not all seem to have the pathway. At the top are the vertebrates only some 400 million years old, and all have the pathway. The arthropods, presumed to have diverged in the early Cambrian and lacking sterol biosynthesis, could well have come, based on all lines of evidence, from the annelids that had failed to acquire the pathway. If this analysis should prove correct, it would imply a totally different origin of the animal line from that of photosynthetic organisms, since the latter, based on the presence of sterol biosynthesis in blue-green algae and on their fossil record, acquired the pathway in the Precambrian about 1–3 billion years ago and maintained it in the familial line, regardless of whether the organism remained or did not remain functionally photosynthetic. No photosynthetic or related organism is known that fails to biosynthesize sterols. Since animals are quite different in this regard, some carrying out biosynthesis and some not, other organisms in the kingdom Nonphotosynthetica might also be expected to be variable in this matter. Fascinatingly, such is indeed the case. Amongst the fungi are the oomycetous genera *Phytophthora* and *Pythium*. In neither of these genera is there sterol biosynthesis, although most fungi do have a complete pathway. The aberrant fungi were first found to lack sterol biosynthesis by Elliott *et al.*, (1964). This has been confirmed more recently by Nes *et al.* (unpublished observations) with *P. cactorum*. However, much as with the arthropods, which require sterols absolutely (cf. Nes and McKean, 1977, and Svoboda *et al.*, 1978), growth of species of *Phytophthora* is enhanced by certain sterols (Elliott, 1968; Nes *et al.*, 1979b), and sterols are an absolute requirement for induction of sexual reproduction (Elliott, 1972). In a related oomycetous genus (*Achlya*), sterol is converted to steroid hormones, which in turn demonstrably induce formation of the sexual apparatus (McMorris, 1978).

It is probable, then, that the sterol requirement of the other oomycetes is for hormone production. Thus, reading between the lines, so to speak, we see a parallel evolution of sexuality and its dependence on steroids. In order to acquire the advantages of sexuality, the sterol-less fungi "learned" to depend on sterol-producing organisms just as yeast "learned" to operate anaerobically by acquisition of exogenous sterol and unsaturated fatty acid. Based on the lack of sterol biosynthesis or requirement in the mycelium, we might hazard a guess that *Phytophthora* and *Pythium* arose very early (Cambrian or Precambrian). Only more recently, then, with the evolution of higher plants, did they develop their present sexuality, in concert with acquisition of infectivity toward tracheophytes, from which they acquire the necessary sterols to be used as hormone precursors. Similarly, it is possible that *Saccharomyces*, having the apparatus for biosynthesis, arose later than *Phytophthora* and *Pythium*, and that the ability of yeast to operate anaerobically developed subsequently in response to the availability of sterols from higher plants.

The foregoing analysis is supported by what we know about the so-called Protista. In the latter category are the protozoa. These organisms, as in *Tetrahymena* and *Paramecium*, lack sterol biosynthesis (Holz and Conner, 1973). The latter depend entirely on exogenous sterol, while the former make the isopentenoid pentacycle, tetrahymanol, in place of a sterol; but the protozoa prefer cholesterol, which strongly inhibits tetrahymanol biosynthesis and replaces tetrahymanol in the membranes. As pointed out by Nes (1974), tetrahymanol mimics the structure of the sterol, but the sterol is recognized as being clearly better. Now, the Protista are believed (Valentine and Campbell, 1975) to have a direct origin from the Precambrian with nothing like the degree of evolution of, say, true animals. As with *Phytophthora* and *Pythium*, the protozoa, at least those mentioned, which are well studied, display a primitive nonphotosynthetic condition, viz., failure to have acquired the sterol pathway. Yet, time has gone on, and they have "learned" to utilize the better material (sterols) obtained from other organisms.

The evidence strongly suggests that (1) the acquisition of the sterol pathway was an extremely important event from both a membranous and a hormonal point of view, (2) this occurred early in the kingdom Photosynthetica and late in the kingdom Nonphotosynthetica, and (3) acquisition was more thorough in plants than in animals. Neither the data nor the analysis should be taken to imply that the four-way division here given (photosynthesis-nonphotosynthesis–plant–animal) is the entire story. The evidence for a separate categorization of, say, the organisms (bacteria) that have membranes of phytyl and biphytanyl ethers strongly suggests they have an independent evolution. It does seem, however, that since the latter organisms are nonphotosynthetic and prokaryotic and lack sterol biosynthesis, they validate the late acquisition of sterol biosynthesis in the Nonphotosynthetica and give credence to a tentative belief in multiple lines of evolution, perhaps from the beginning, in which sterol biosynthesis was acquired very early only in a certain group (which excludes the photosynthetic bacteria) of organisms, viz., the blue-green algae.

We shall now return to the detail of the reduction–alkylation bifurcation that discriminates between plants and animals. The mechanism of the two reactions has been the subject of extensive investigation. Reduction of the Δ^{24} bond proceeds by attack of a proton to C-24 and of a hydride ion from NADH to C-25 (Wilton *et al.*, 1970; Greig *et al.*, 1971). The mechanism (Figure 23) of alkylation is very similar (Castle *et al.*, 1963, 1967). A carbon atom from S-adenosylmethionine acting as if it were a methyl carbonium ion, the carbon analog of a proton, also attacks C-24. Enzymological evidence for this similarity in mechanisms is that both are inhibited by the same substance, triparanol (Malhotra and Nes, 1971). Inhibition occurs in cell-free systems, suggesting a competitive mechanism at the enzyme level. In algal systems,

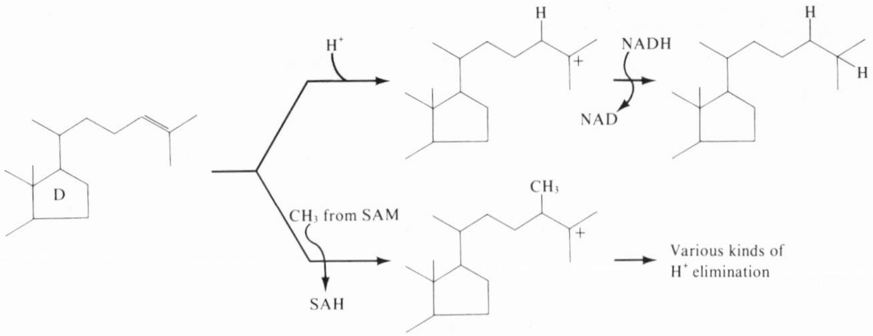

Figure 23. Similarity of mechanism of the reduction–alkylation bifurcation.

triparanol disrupts sterol biosynthesis multiply, probably by interfering with the endoplastic reticulum on which all reactions occur after squalene. The difference in the two routes (reduction vs. alkylation), aside from the nature of the electrophile attacking, lies in what happens after formation of the C-25 carbonium ion. Reduction, so far as is known in both plants and animals, occurs exclusively with hydride attack on the C-25 carbonium ion. Conversely, in alkylation this never is known to happen, although it is possible (Castle *et al.*, 1963). The alkylated carbonium ion of Figure 23 always loses a proton, yielding a double bond. One of the possibilities is to lose a proton from the incoming methyl group with hydride transfer from C-24 to C-25 (Raab *et al.*, 1968), producing a $\Delta^{24(28)}$ bond (24-methylenesterol). As first appreciated in this laboratory (Castle *et al.*, 1963), the 24-methylene group permits a second alkylation by C_1 transfer to the methylene group (C-28) (Castle *et al.*, 1967) and accounts in principle for the existence of the two major types of phytosterol (24-C_1 and 24-C_2). However, the detail of these processes constitute a polyfurcation in the pathway that has extensive phylogenetic significance. We will examine each of the routes separately.

A bifurcation that is immediately obvious from the top line of Figure 24 and in many ways the simplest of the divergencies in the pathway is the orientation of C-29 in the 24-ethylidenesterol following the second C_1 transfer from *S*-adenosylmethionine. The *cis* or E sterol (C-29 on the left) is formed, so far as is known, only in brown algae (Phaeophyta), where it undergoes no further metabolism. The dominant sterol of the phaeophyta has this E-ethylidene structure in the Δ^5 series (fucosterol). The Z-ethylidenesterols, e.g., isofucosterol, are now known to be formed widely in tracheophytes, where these sterols are also converted to 24-ethylsterols as first shown in this laboratory (Van Aller *et al.*, 1969). There thus seems to be a phylogenetic and metabolic correlation between the configuration of the ethylidene group and where it is formed and whether or not it is reduced (Van Aller *et al.*, 1968). The general consensus among investigators at the present time adheres to the ideas implicit in the foregoing discussion, viz., that in tracheophytes only the Z-isomer is produced and that the E isomer is confined to the brown algae. Further work on this relationship would be worthwhile, because it has been reported that exogenous fucosterol will proceed to 24-ethylsterols in green algae (Tsai and Patterson, 1976) and that fucosterol is an intermediate between sitosterol and cholesterol in insects (Svoboda *et al.*, 1971; Allais and Barbier, 1971).

Another bifurcation obvious from Figure 24 is whether the C-25 carbonium ion with a 24-methyl group eliminates a proton from C-28 or C-27. The former choice leads, as just discussed, to introduction of a $\Delta^{24(28)}$ bond, while the latter yields a $\Delta^{25(27)}$ sterol. These two routes are extremely well documented (Nes, 1977; Nes and McKean, 1977, Goad and Goodwin, 1972). The $\Delta^{24(28)}$ route through the 24-methylenesterol is always necessary for the second alkylation. Conversely, the $\Delta^{25(27)}$ route at the 24-C_1 stage precludes a second alkylation and can only lead to 24-C_1 sterols.

24-C_1 Sterols will then be seen to have two routes open to them. In the one

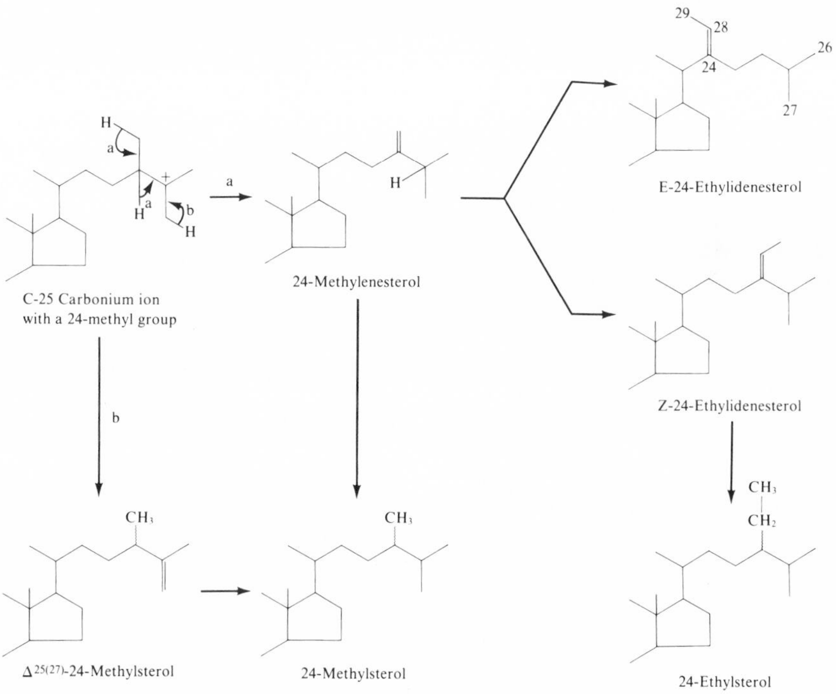

Figure 24. Origin of phytosterols. Figure ignores stereochemistry at C-24 and intermediacy of various kinds of unsaturation.

case via elimination from C-28 a 24-methylenesterol is produced. In the other case elimination from C-27 yields a 24-methylsterol with a $\Delta^{25(27)}$ bond. Now, in the latter case (and only in the latter case) a question of chirality arises at C-24, because this carbon atom is tetrahedral, unlike the trigonal condition in the 24-methylenesterols. In all cases for which stereochemical information is available, e.g., in cyclolaudenol, the methyl group (of the $\Delta^{25(27)}$ sterol) is in back (β). We therefore tentatively assume that all $\Delta^{25(27)}$-24-methylsterols have the β configuration at C-24. Since cyclolaudenol is isolable along with 24β-methylcholesterol and other 24β-methylsterols, e.g., brassicasterol, from tracheophytes we assume that the 24β-methylcholesterol arises by the $\Delta^{25(27)}$ route through reduction of the $\Delta^{25(27)}$ bond (McKean and Nes, 1977a). This is supported by labelling evidence. For similar reasons, the origin of 24β-methylsterols in green algae appears to proceed via the $\Delta^{25(27)}$ route. However, in fungi and golden algae the 24β-methylsterols arise by reduction of 24-methylenesterols. (For a detailed discussion of the evidence, see Nes, 1977.) Particularly important is the number of deuterium atoms transferred from CD_3-S-adenosylmethionine. In the $\Delta^{24(28)}$ route only two will appear in the methyl group due to the intermediacy of the methylene group. The third H atom would be derived from water as a proton during reduction (presumably by H^+ and NADH, as is the demonstrable case at the Δ^{24} bond). In the $\Delta^{25(27)}$ route all three deuterium atoms must be found in the methyl group, since reduction occurs elsewhere (C-25 and C-27). All fungi, including a lichen mycobiont, and golden algae incorporate only two D atoms, but all green algae incorporate three D atoms. In all of these organisms, (except diatoms, to be discussed subsequently) the 24-C_1 component has the β configuration, and both the configuration and the route are independent of other structural features, e.g., whether the sterol is in the Δ^5, $\Delta^{5,7}$, Δ^7, or Δ^{22} series or whether the product retains the nuclear methyl groups (eburicoic acid, etc.). This means (1) that exactly the same compound can arise in two different organisms by different routes (as illustrated in Figure 25) and (2) that the mechanism of alkylation has a taxonomic and therefore at some level a phylogenetic relationship. Ergosterol, for instance, is formed in both fungi and green algae, but it arises by the $\Delta^{24(28)}$ route in fungi and the $\Delta^{25(27)}$ route in green algae. It would be attractive to associate this with the presence or absence of photosynthesis, but the situation with golden algae proves this cannot be done. Which of the two routes is used is independent of the photosynthetic question and independent of whether cycloartenol or lanosterol is the product of cyclization of epoxysqualene. Both green and golden algae, each differently forming 24β-methylsterols, use the same mode of cyclization through cycloartenol, whereas fungi and golden algae both have the same route to 24β-methylsterols and different modes of cyclization of epoxysqualene (the lanosterol and cycloartenol routes, respectively).

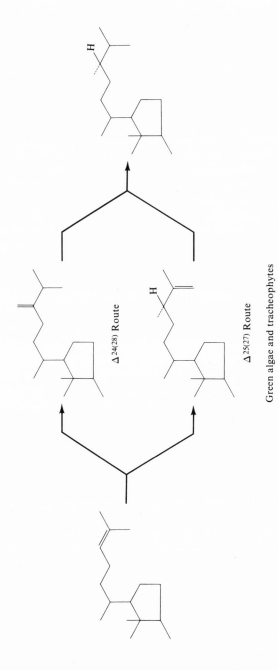

Figure 25. Bifurcated route to 24β-methylsterols. Diatoms unlike other golden algae reduce the $\Delta^{24(28)}$ bond to give a 24α-methylsterol.

The use of the $\Delta^{24(28)}$ vs. $\Delta^{25(27)}$ routes for the formation of 24-methylsterols actually transcends the configurational question at C-24, although the route and configuration clearly, as just discussed, are not totally unrelated. The problem revolves around (1) the side of C-24 attacked by the incoming methyl group from S-adenosylmethionine, and (2) whether C-24 is tetrahedral or trigonal when reduction of the double bond occurs. If, as seems to be the case, methylation with introduction of a $\Delta^{25(27)}$ bond always occurs from the back or β side of C-24 and the $\Delta^{25(27)}$ bond subsequently suffers reduction, the configuration is set (β) by C_1 transfer. However, in the $\Delta^{24(25)}$ route the configuration is set by the reduction, and the side attacked by the reductant appears to depend on the organism and not on the substrate or product. Thus, reduction of 24-methylenesterols in fungi leads to 24β-methylsterols, but reduction of 24-methylenesterols in golden algae appears to lead to either 24β- or 24α-methylsterols. Brassicasterol arises in *Ochromonas malhamensis* (a phytoflagellate) and 24-epibrassicasterol (diatomsterol) is formed in *Phaeodactylum tricornutum* (a diatom). In both organisms only two D atoms are transferred, despite opposite configurations (Rubinstein and Goad, 1974). The phytoflagellates and diatoms are both in the division Chrysophyta but represent different classes. The phytoflagellates are class Chrysophyceae and the diatoms class Bacillariophyceae. The third class of the Chrysophyta is constituted by the Xanthophyceae. These algae are like the phytoflagellates in their biosynthesis, leaving the diatoms singular not only in the division but amongst all algae. Only the diatoms have (and have exclusively) 24α-alkylsterols (which occur at both 24-C_1 and 24-C_2 levels).

Unlike the algae and fungi, which seem to operate the route to 24-methylsterols in one way in a given organism, tracheophytes use multiple routes. 24-Methylcholesterol, for instance, has been shown to be an epimeric mixture in main-line plants (Nes *et al.*, 1976b). The 24α-methyl component (campesterol) dominates (ca. 2:1) over the 24β-methyl isomer (dihydro-brassicasterol). The present evidence (McKean and Nes, 1977a), which needs to be expanded, suggests that the 24α-methyl route arises via a 24-methylenesterol ($\Delta^{24(28)}$) and the 24β-methyl component by the $\Delta^{25(27)}$ route. Unfortunately, the $\Delta^{24(28)}$ route could be more complicated than simple reduction of the $\Delta^{24(28)}$ bond. The presence of 24-methyldesmosterol (the $\Delta^{24(25)}$ isomer of 24-methylenecholesterol) in a tracheophyte (*Withania somnifera*) raises the possibility that the $\Delta^{24(28)}$ bond is isomerized to the more stable $\Delta^{24(25)}$ position and that the latter double bond is what undergoes reduction (Goad *et al.*, 1974). This would not change the number of D atoms incorporated, making assessment of the problem difficult.

When a second C_1 transfer occurs to a 24-methylenesterol, the same sort of polyfurcation exists as outlined above for the first C_1 transfer, with the added problem of *cis-trans* isomerism. Methylation of C-28 can produce either a $\Delta^{24(28)}$ sterol (E- or Z-24-ethylidenesterol) or a $\Delta^{25(27)}$-24-ethylsterol. Both routes occur, and in the algae what happens corresponds to the routes for the first C_1 transfer. The green algae, for instance, incorporate five D atoms into the ethyl group ($\Delta^{25(27)}$ route), and the golden algae incorporate four ($\Delta^{24(28)}$ route). $\Delta^{25(27)}$-24β-Ethylsterols are also reduced demonstrably at the $\Delta^{25(27)}$ bond in green algae (Largeau *et al.*, 1977a). As at the 24-C_1 stage, exactly the same compound, e.g., 24β-ethylcholesterol (clionasterol), can arise by the two different routes (Figure 26). In diatoms, although labelling data are not available at the 24-C_2 stage, the 24-ethylsterol is of the α configuration as it is at the 24-C_1 stage, suggesting here too the same route ($\Delta^{24(28)}$) at both levels of C_1 transfer. In tracheophytes, two families (Verbenaceae and Crassulaceae) form exclusively 24β-ethylsterols bearing a $\Delta^{25(27)}$ bond (cf. Nes *et al.*, 1977a). It seems clear that these plants use the $\Delta^{25(27)}$ route.

The 24α-ethyl pathway in tracheophytes proceeds by a 24-ethylidene intermediate, since isofucosterol is converted experimentally to sitosterol and four D atoms are incorporated, respectively, in *Pinus pinea* (a gymnosperm) and *Hordeum vulgare* (an angiosperm). However, as in the 24α-methyl case, there is reason to believe in an intermediate isomerization of the $\Delta^{24(28)}$ sterol to a sterol with a $\Delta^{24(25)}$ bond prior to reduction (Largeau *et al.*, 1977b; McKean and Nes, 1977a). Although there has been difficulty in proving the existence of the intermediate or showing that administered $\Delta^{24(25)}$ sterol will produce the 24α-ethylsterol, there is no question that the 25-H atom (originally at C-24) is lost. This cannot be explained well by a route other than through a $\Delta^{24(25)}$ intermediate. The best judgment is that 24α-alkylsterols are probably always produced by this route in tracheophytes (Figure 27). Whether the diatoms directly reduce the $\Delta^{24(28)}$ bond or also proceed with loss of the 25-H atom remains to be seen.

In summary, sterol biosynthesis seems to have arisen in the Kingdom Nonphotosynthetica variously and probably for the most part after the Cambrian. Nonphotosynthetic organisms diverged into two groups: those with and those without a pathway to sterols. Those that developed a nervous system (animals) evolved a requirement for the unalkylated side chain in sterols regardless of the exogenous or endogenous source of sterol, and when *de novo* biosynthesis arose, only a reductase for the Δ^{24} bond became available, presumably in order to preserve the functional integrity of the nervous system. That part of the nonphotosynthetic line which did not develop a nervous system (fungi) moved evolutionarily in both directions with respect to the reduction–alkylation bifurcation, since no restriction of a nervous system was present. An array of sterols with no alkyl group, a 24-methyl, or a 24-ethyl group came to be biosynthesized. No animals are known in the kingdom Photosynthetica, so all of the organisms in this kingdom were

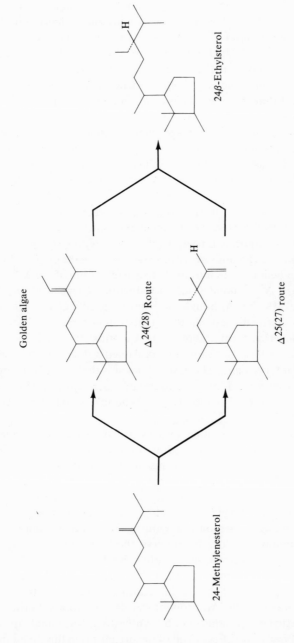

Figure 26. Bifurcated route to 24β-ethylsterols. Diatoms, unlike the other golden algae, produce 24α-ethylsterols, which presumably arise by the Δ²⁴⁽²⁸⁾ route as do sterols at the 24-C₁ level.

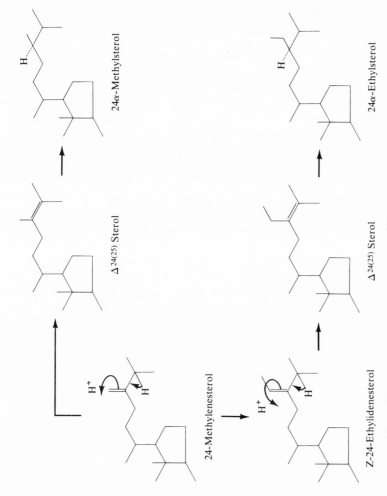

Figure 27. Route to 24α-alkylsterols. (The $\Delta^{24(25)}$ intermediates are not fully proven.)

free to evolve along with the fungi toward either alkylation or reduction. Some of the photosynthetic plants, notably among the red algae but also in a few cases in organisms as highly developed as tracheophytes, form primarily or exclusively cholesterol or other sterols with an unalkylated side chain, while most, whether algal or vascular, alkylate the side chain. The mechanism of alkylation, constituting a sequential, subordinate polyfurcation relative to the reduction-alkylation bifurcation, varies with the division (phylum) and within the division, and the mechanism determines what sterols can be present. The $\Delta^{25(27)}$ route can lead only to 24β-alkylsterols. When this is the only mechanism of alkylation available, as in the green algae and a few tracheophytes, only 24β-alkylsterols are present. The $\Delta^{24(28)}$ route is amenable to formation of either 24α- or 24β-alkylsterols, and it is utilized variously in the plant kingdom without regard for the degree of cellular differentiation or the presence or absence of photosynthesis. Golden algae produce both 24α- and 24β-alkylsterols by this route. Fungi seem to use it only for 24β-alkylsterols, and tracheophytes probably use it only for 24α-alkylsterols. Unlike algae, tracheophytes, even in the same plant, have both the $\Delta^{24(28)}$ and $\Delta^{25(27)}$ routes available, but not always.

It is very difficult to use the extensive information we have to deduce phylogenetic lines within the plant kingdom owing to the variable (statistical) nature of what happens. For instance, an attractive photosynthetic line on other grounds would be blue-green algae to red algae to green algae to mosses to gymnosperms to Magnoliales or Ranales and then on to higher angiosperms. Unfortunately, there is no direct line apparent with respect to the reduction-alkylation bifurcation and its subordinate polyfurcation. Blue-green algae primarily alkylate; red algae primarily reduce; green algae primarily alkylate and use only the $\Delta^{25(27)}$ route; Magnoliales and Ranales, both in the main line, primarily alkylate, giving 24α-ethylcholesterol as the dominant sterol, presumably by the $\Delta^{24(28)}$ route; while some still more highly developed plants, e.g., the Crassulaceae and Verbenaceae, use only the $\Delta^{25(27)}$ route to 24β-ethylsterols, and others (most) make no 24β-ethylsterols, forming only the α configuration at the 24-C_2 level by the $\Delta^{24(28)}$ route as do Magnoliales and Ranales. There appears to be a mix composed of functional requirements for a given sterol and the manner in which the sterol is made. The two components of the mix seem to have evolved independently. This is illustrated on the one hand in the case of the same compound, ergosterol, formed by different routes in fungi and algae as well as by the use of the same route for different sterols in different classes of golden algae, and by variation of the routes and sterols among the tracheophytes. An alternative accounting for the pathways would be to assume some such thing as that main-line tracheophytes (with two routes to 24-alkylsterols) arose from a symbiosis of diatoms (one route) and green algae (the other route), but this is absurd. In the absence of any consistency, we feel it more likely that large-scale genetic lines do not exist with respect to the alkylation. Ways of carrying out alkylation seem always to have been available and seem to have been introduced in a given case for reasons not apparent except for ecological-functional interactions. The importance of functionality seems especially clear with animals in which all routes of alkylation have been excluded. This is the more remarkable from a statistical point of view, when it is remembered that more than one route of alkylation is possible. If animals had developed, say, the apparatus for conversion of $\Delta^{24(28)}$ sterols to 24-methyl- or 24-ethylsterols but simply never acquired (or lost) the C_1 transferase necessary to give 24-methylenesterols, then they should be able to reduce or alkylate 24-methylenecholesterol, which is not so in the case of mammals (Nes *et al.*, 1973). Similarly, they are never known to possess either 24-methylenesterols or 24-methyl- or ethylsterols (by biosynthesis), so they must have the apparatus for neither the first transfer nor the subsequent parts of the process. This extensive restriction, documentable in other kinds of processes, e.g., in the kinds of epoxysqualene cyclase present, strongly suggests either a feedback of function on biosynthesis or a previously unrecognized organizational principle.

3. Primitive Routes

Some of the criteria by which primitive routes might be judged are (1) whether the pathway is completed, (2) whether a low- or high-energy route is used, (3) whether the pathway is aerobic or anaerobic or the extent to which it is aerobic, and (4) chirality.

As to point (3), from Chapter 5 it will have been seen that the supposition of an early anaerobic environment on the earth is at best speculative, so the reason for assuming that an anaerobic route is a primitive one cannot be taken as compelling. Nevertheless, Rohmer *et al.* (1979) has suggested that the existence of 3-desoxyhopanes (pentacyclic triterpenoids derived by anaerobic cyclization of squalene) in many prokaryotes is illustrative of a primitive, anaerobic pathway that yields a material able to mimic the functional role of sterols in membranes. Such a functional role for tetrahymanol and the carotenols was proposed earlier by Nes (1974), who also called attention to the possible relationship between the anaerobic route of biosynthesis of tetra-hymanol and primitive character. The question of energetics is related to these ideas. In the simplest sense one would like to believe that energetically less expensive processes occurred before the more expensive ones. This is especially attractive in an entropy sense with regard to three-dimensional orientations. For instance, Nes *et al.* (1967) pointed out that, since the formation of sterols proceeds through an energetically less favorable conformation than do triterpenoids, the latter (β-amyrin, hopanoids, euphoids) should be more primitive than the former. A similar argument was made subsequently by Rohmer *et al.* (1979). Among the problems with this is that the distribution of triterpenoids does not correlate well with such ideas. The triterpenoids are found in prokaryotes, but only in some of them, and triterpenoids seem to be very rare among eukaryotic algae. In higher plants, however, they are frequently found, yet they are not found in animals. This suggests a functional reason for the selection of a sterol or triterpenoid rather than an energetic one. The evolutionary idea was first suggested experimentally by the sequential appearance of squalene, β-amyrin, and sterols in germinating peas (Baisted *et al.*, 1962; Capstack *et al.*, 1962; Capstack *et al.*, 1965; Nes *et al.*, 1967). The data are in agreement with the dual postulate of anaerobiosis and a low-energy process being primitive, if we additionally assume evolutionary recapitulation, but subsequent work on germinating pines (McKean and Nes, 1977b) showed that both maternal and embryonic tissue of the seed accumulate squalene during early ontogeny and that neither accumulates a pentacycle. The sequential formation of these compounds is probably interpreted better in terms of function than of timing in evolution. Furthermore, if we were to take the energetic idea in detail, we would also have to conclude that the cycloartenol pathway is more advanced than the lanosterol pathway. Cycloartenol is both theoretically and experimentally of higher energy than lanosterol. Again, occurrence data do not agree. The evidence we have is that the cycloartenol pathway is as old as or older than the lanosterol pathway and that both widely coexist today.

Chirality has always been an intriguing possibility by which evolution might be judged. For a while it seemed that all earthly creatures, with or without a nervous system, have the same chiral principles operative. Indeed, many facts are in agreement with this. Among them is that regardless of evolutionary position, epoxysqualene can be cyclized in the same chiral fashion such that only one (and the same) enantiomer of sterols is formed whether in algae, fungi, or animals. This idea, though, was shattered as an absolute rule by the discovery of D-amino acids in bacteria. Equally or more important was the discovery of enantiomeric terpenes, e.g., citronellol, linalool, limonene, and pinene. (For a key to the literature, see Karrer, 1958.) Depending on the compound, they are found as one enantiomer in one plant, the other in another plant, and on occasion as the racemate in still other cases. As pointed out in greater detail by Nes *et al.* (1967) and Nes and McKean (1977), the frequency of occurrence of enantiomeric forms seems to be inversely related to the extent of polymerization of the isopentenoid unit (I). Thus, there appears to be increasing stereoselectivity as the complexity of the molecule increases from the I_2 to the I_3 to the I_4 to the head-to-head dimers of the structure I_3–I_3 and I_4–I_4 (Nes *et al.*, 1967). While it is attractive to believe that this represents some evolutionary ordering in time, with the simpler, racemic molecules being the earliest, the occurrence data lend no support. All of the compounds, whether simple, complex, enantiomeric, etc., are found in tracheophytes, which are among the most highly differentiated and recent organisms on earth. If we turn the problem from enantiomerism to diastereoisomerism, much the same things are found. Diastereoisomers exist in different organisms as well as in the same organism. For instance, euphol and lanosterol are diastereoisomers, and they coexist in plants belonging to the family Euphorbiaceae. 24α- and 24β-Methylcholesterol also commonly coexist in tracheophytes, whereas 24α- or 24β-alkylsterols of various kinds can be the exclusive component of a given plant both among the algae and the tracheophytes. As pointed out before in this chapter, the statistics of occurrence are what seem to be more meaningful. Very few tracheophytes make only 24β-alkylsterols, while very few algae make 24α-alkylsterols. Here there seems to be an association with the degree of cellular differentiation. The 24α-alkyl group, especially at the 24-C_2 stage, seems to be better for more complex plants than the epimeric condition. Interestingly, among the Cucurbitaceae, the 24-ethylsterols of seeds have the 24β configuration, while the corresponding sterols of the mature plant have the 24α configuration (Nes *et al.*, 1977a). Does this represent evolutionary recapitulation, which is a very attractive explanation, or is it a functional matter? A firm answer will have to await further work and thought. Unlike the situation at C-24, the sterols of all examined algae, fungi, tracheophytes, and animals have a single configuration at C-20 (Nes *et al.*, 1977b), but to confuse the evolutionary picture epimers at C-20 exist in the euphoid series. More work again will clearly have to be done to understand the full significance of the chiral problem. There seems to be no absolute rule which tells us, for instance, that an isopentenoid chain must always be arranged in a particular chiral conformation at the time it undergoes cyclization on this planet, any more than there is an absolute rule telling us only L-amino acids will be made. It does seem to be true, though, that similar chirality exists in *most* organisms. Moreover, the factors determining chirality seem to have a very strong organizational component. By this we mean the more complex the system the more likely it is to find similarity in chirality. This component of complexity also seems to be subdividable into molecular and organismic categories. If the molecule or the organism is simple, there is a greater chance of having enantiomeric processes than if one or both are complex. The data cited above agree with this principle. Both D- and L-amino acids occur in prokaryotes, but only the L-enantiomers in eukaryotes. Simple terpenes but not sterols exist as enantiomers.

As for whether or not the sterol pathway is completed, two questions arise. One is qualitative, viz., whether all steps occur from acetate to sterol, and the other is a quantitative question about the kinetics. The suggestion has been made (Nes *et al.*, 1967) that isopentenoid biosynthesis could be viewed as a molecular tree in which the evolutionary process began with acetate and proceeded in time to make the more complicated molecules by adding steps and, at appropriate places, branching in new directions. The simpler cells, assuming they represent an earlier time, should then have less of the isopentenoid pathway than the more complicated ones. The facts lend only some support. For example, blue-green algae have a full pathway to sterols, but insects have none. Moreover, there is just no simple addition of pieces of the pathway as the evolutionary ladder is climbed. Some bacteria have the pathway complete to squalene, while among flatworms, which are eukaryotes, the pathway diverges prior to squalene with the formation of 2-*cis*-6-*trans*-farnesol instead of all-*trans*-farnesol. To make matters worse, in animals lacking the pathway, additions are made at the end. This is seen, among other examples, in the dealkylation of sitosterol to cholesterol in insects, even though the insect cannot make sitosterol. Similarly, cholesterol is converted to cholestatrienol in protozoa, which cannot make cholesterol. In terms of other isopentenoid pathways, there are deletions at the beginning, as in the requirement of *Lactobacillus acidophilus* for an exogenous source of mevalonate, which is then used to make carbohydrate carrier lipid. These additions and deletions cause us a great deal of trouble in trying to make a case for a building-up principle in which parts of the pathway are added sequentially. The best example of a positive relationship is among the bacteria, where some have no squalene, some have squalene, some have cyclic intermediates, and some have sterols. This is exactly what we would like to see, but it is both restricted to nonphotosynthetic prokaryotes and manifested only in a statistical sense. Nor do the data correlate well with the supposition of a loss of the pathway in those organisms that lack all or part of it. If this were so, we should frequently find insects, for instance, with pieces of the pathway missing in the middle, at the beginning, or at the end, but from what can be seen so far, such a random distribution does not exist. In terms of the steroid pathway, rather than randomness we see a positive relationship between additions to the sequence and the eating habits of these animals. The carnivorous insects do not have a dealkylating mechanism for phytosterols, while phytophagous insects do.

The other possible way of assessing primitiveness in the pathway, i.e., by a kinetic parameter, may have more promise. Blue-green algae seem to operate the pathway without the accumulation of intermediates except in some cases at the end where, for instance, Δ^7 sterols accumulate, which is a drawback to the idea. Nevertheless, the strong accumulation of intermediates with nuclear methyl groups in some mosses and the complex distribution of Δ^5, $\Delta^{5,7}$, and Δ^5 sterols in both prokaryotic and eukaryotic algae compared to the kinetically complete pathway of animals and most tracheophytes, which have almost only Δ^5 sterols, is in keeping with the general idea of an evolution of kinetics such that as time proceeds intermediates are kept more and more precisely to a very small, steady-state concentration.

4. Sequence

There are three aspects to an analysis of the sequence of events in a pathway (Nes, 1971). In one of these we must consider the molecule itself. Certain reactions must occur prior to others for intrinsic chemical reasons. For instance, during C_5 polymerization, C_5 must come before C_{10}, etc. We can also look at this problem in a more sophisticated way. Thus, it is not possible to break a saturated carbon–carbon bond without first having appropriate functionalization in terms of electronics. An example is found in the aerobic hydroxylation of the methyl groups on C-4 prior to their removal as CO_2. That is, in the conversion, say, of lanosterol to 4-desmethyllanosterol plus CO_2 in the absence of an electron donor such as homocysteine, the sequence must be oxidation followed by rupture of the C–C bond rather than the reverse. Similarly, the addition of a carbon atom at C-24 from S-adenosyl-methionine requires the prior presence or introduction of a Δ^{24} bond. Stated more generally, for a given type of reaction to occur some other reaction must precede it in order to establish appropriate electronic, energetic, or steric conditions in the molecule. If we then know what the reactions are in a sequence, we can predict the order in which they must occur. There is, however, a second level of sequence in which no such prediction can be made, because the reactions involved are at different places in the molecule and completely unrelated to one another. Thus, there is no intrinsic reason why demethylation at C-4 should occur before or after methylation at C-24. The control for the sequence in this case has to lie in something other than the molecule itself. The factors involved would include the Michaelis constants of the enzymes, ordering of the enzymes on a membrane, and binding constants of any carrier proteins which have to play a role. Third, in order to analyze the sequence, we have to consider cases in which the same result can be achieved either by a single reaction or by a set of reactions. An example is found in the formation of a Δ^8 sterol. Epoxysqualene can proceed directly to lanosterol by cyclization, but it may also cyclize instead to cycloartenol, which then proceeds by a new reaction (isomerization) to lanosterol. These three aspects of the sequence raise a number of opportunities to examine evolution and phylogenetics.

In the first place, if life evolved in an orderly, stepwise manner, one could imagine that the earliest reactions to evolve would have been the ones required intrinsically early in a pathway. This would imply that the enzymes for formation of geranyl pyrophosphate (GPP) are older than those required in the conversion of GPP to squalene, which are in turn older than squalene epoxidase, which would be older than epoxysqualene cyclase, etc. Some evidence for this exists in that, for instance, there is every reason to believe that nearly all cells, whether prokaryotic or eukaryotic, have the capacity to make GPP, while the ability to make sterols is much more restricted. One could follow this idea through in greater detail, but not yet enough is known about interacting parameters to place it in the realm of a proven principle. However, it may be reasonable in some cases to say as a tentative postulate that an organism that carries the pathway only to some point and not further is more primitive in this respect than an organism that has the capacity to operate the entire pathway. Thus, some of the bacteria that make only squalene might be considered more primitive than those that cyclize it; a cell that makes only 24-dehydrocholesterol (desmosterol) could be considered less advanced than one that has the Δ^{24} reductase or the Δ^{24} alkylase and, therefore, makes cholesterol or 24-alkylcholesterol, respectively; and the ability to make only sterol and not to further metabolize it might be construed to be more primitive than the capacity not only to make sterol but to convert it, say, to bile acid or hormone.

There are attractive correlations between such ideas and other parameters. In the bile acid case, for instance, as Haslewood (1967) has pointed out (see also Anderson *et al.*, 1974; Ikawa and Tammar, 1976; and Kellogg, 1975) the progression from cholesterol to bile alcohols with all 27 carbon atoms of cholesterol to C_{27} acids to C_{24} acids can roughly be seen to parallel vertebrate evolutionary development. Invertebrates do not seem to metabolize cholesterol at all to bile salts. The crab, *Cancer pagurus*, for instance, uses the fatty acid derivatives, decanoyl sarcosyl taurate, instead of a steroid as an emulsifying agent in its gastrointestinal system. However, primitive craniate chordates have steroidal bile alcohols. The hagfish, for instance, makes a disulfate of myxinol (5β-cholestane-$3\beta,7\alpha,16\alpha,27$-tetrol) bearing 27 C atoms, and the lamprey makes the C_{24} alcohol 3α, 7α, 12α, 24-tetrahydroxycholane. The sharks and other "cartilaginous" fish (Chontrichthyes) make a sulfate of a C_{27} alcohol, scymnol (5β-cholestane-$3\alpha,7\alpha,12\alpha,24\xi,26,27$-hexol); and in the bony fish (Osteichthyes), while some have bile alcohols, bile acids make their appearance, as in the case of the sturgeons, which have the C_{24} acid cholic acid ($3\alpha,7\alpha,12\alpha$-trihydroxy-5β-cholan-24-oic acid). Both bile alcohols and bile acids are also found in amphibians, but only the acids have been encountered in reptiles, birds, and mammals. Furthermore, reptiles have both C_{24} and C_{27} acids, but the pathway is complete to C_{24} acids, e.g., cholic acid, in birds and mammals. When examined in greater detail, the original 3β-hydroxy and pseudo-A/D-*trans* configuration of the Δ^5 sterol appear to be more frequent lower in the vertebrate scale than in the higher regions. Mammals have only 3α-hydroxy-A/B-*cis* acids, while among the fish one finds the more primitive configurations. For instance, the 3β-hydroxy-A/B-*trans*-alcohol 5α-latimerol (5α-cholestane-$3\beta,7\alpha,12\alpha,26,27$-pentol) is found in the most primitive of the bony fish (the coelacantha). It would of course take the evolution of appropriate enzymes to invert C-3. The question of C-5 is more sophisticated. The A/B-*trans* configuration (5α) is both the more stable and the easier to achieve mechanistically from a Δ^5 steroid (attack of H donor from the less hindered side). Consequently, we would expect it (and find it) to arise first from a Δ^5 sterol. In this connection, it is worth noting that the 3β-hydroxy configuration, which being equatorial happens to be the more stable, is a necessary consequence of the mechanism of cyclization. In the spiral conformation of epoxysqualene leading to the A/B-*trans* series, the oxygen atom must be on the side opposite to C-5, for stereoelectronic reasons (maintenance of *trans* reactions). A 3α-hydroxy-5α-steroid could not reasonably be obtained by redistribution of electrons induced by protonation of the oxygen atom. The mechanism, rather than the final state, governs the outcome. More interesting is that the mechanism just happens to give the more stable state and that all subsequent processes maintain and use this state except for highly advanced cases, e.g., 3α-bile acids and hormone metabolites in the higher vertebrates. It has to be mentioned, however, that once the A/B-*cis* system is reached, the 3α-hydroxy group becomes the equatorial and hence the more stable configuration in the (more stable) chair form of ring A. Herein lies the complication and satisfaction of our evolutionary considerations. A higher functional reason, emulsification of dietary material, dictates that a structure with the A/B-*cis* configuration should be formed. This is demonstrably less stable than the A/B-*trans* analog. It (*cis*) arises statistically only in the chronologically higher forms of life, but when it does arise, we find thermodynamics pulls the events toward production of the more stable system (3α-hydroxy) in the more highly evolved organisms (mammals). In the bile acids, then, we see the pull of opposing forces operating in a chronological milieu. One of the forces appears to be the necessity, in terms of fulfillment, for the development of the perfect emulsifying agent, but this takes a pathway; the pathway must be developed stepwise and therefore in time; and all the while mechanisms and thermodynamics must be contended with. The evolution of a cholic acid conjugate (a C_{24}-3α-OH-A/B-*cis*-carboxylic acid ester of a polar alcohol) in human beings can be viewed as having occurred under the influence of (1) a goal-directed component (function as an emulsifying agent) operating against the thermodynamics of the relative stabilities of A/B-*trans* and A/B-*cis* systems, (2) an energetic component (rearward attack to give an equatorial isomer) operating against or at least independently of the functional component, and (3) still another independent component in which the goal can only be reached stepwise and consequently in a chronological sequence.

The second idea presented at the beginning of this subsection also has some known consequences. It has to do with events in a molecule which are intrinsically independent. Taking the bile acid example, at what stage of the pathway should a 12α-hydroxy group be introduced? Situated as it is in the middle of the molecule away from ring A and the side chain, there is no intrinsic reason why it should be introduced at any particular stage. Where it is introduced in terms of the sequence could be dictated by other factors, which means that the biosynthesis of cholic acid, a 12α-hydroxy compound, could reasonably vary from animal to animal in terms of the exacting sequence of events. In a related case, for which much more information is available, the position in the sequence at which C_1 transfer to C-24 occurs in the formation of 24-alkylsterols should have no relationship to the events in rings A and B. Let us see what the facts are. In agreement with the foregoing proposition, C_1 transfer is actually known to be independent of many other events and correlative instead with organismic type. Before outlining examples of this, we must, however, call attention to one glaring exception.

Based on what we have said, C_1 transfer could occur randomly. Actually, the evidence is that alkylation never occurs prior to the cyclization of epoxysqualene. The main reason for believing this is that no alkylsqualene has ever been encountered in any organism, despite many reports on squalene and its reduced derivatives. A precise example is found in germinating pea seeds, which more or less exclusively make 24-alkylated sterols, yet squalene and not 24-alkylsqualene is found in the seeds. Similarly, only squalene is formed in a very different organism, yeast, which also makes 24-alkylsterols. There is no intrinsic chemical reason why this should be. The required double bond for alkylation is present in squalene as it is in its tetracyclic products. We have to attribute this interesting fact to an organizing principle that as yet escapes us but presumably is goal-directed. Alkylation only occurs after squalene is cyclized. Moreover, alkylation occurs only in the tetracyclic series, only in the 3-hydroxy series, and only in the side chain, despite the fact that double bonds are available at many positions in the ring structures of both tetracycles and pentacycles. There are hundreds of molecular examples to document this throughout the sweep of organismic evolutionary development. The press of function is nowhere better seen than in the discrimination made by living systems in their choices for alkylation of double bonds.

Accepting empirically that alkylation can occur only after cyclization of epoxysqualene, we find as expected that there is no absolute choice as to whether an extra C atom should be added immediately after cyclization or at some later biosynthetic stage. Some well-documented examples are as follows. Most higher photosynthetic plants investigated introduce a C_1 group immediately after cyclization. Thus, 24-methylenecycloartanol (or perhaps in some cases cyclolaudenol) appears to be the prime metabolite of cycloartenol, but not in the euphorbs. In the family Euphorbiaceae the first reaction following cyclization appears to be isomerization to lanosterol. Lanosterol (and 24-dihydrolanosterol) along with cycloartenol are well-established products in euphorbs, while 24-methylenecycloartanol is well established both as present and as a biosynthetic intermediate in other plants. Despite this, both euphorbs and the other plants make the same 4-desmethylsterol, sitosterol. Here we have commonality in two different plants with respect to the precursor and product (cycloartenol and sitosterol) but differences in the sequencing of events in between. In the euphorbs the three-membered 9,19-cyclogrouping is opened before the Δ^{24} bond is alkylated, while in most of the other plants studied, the reverse occurs. A similar reversal of events is found in the fungi. In *Saccharomyces cerevisiae*, which cyclizes epoxysqualene to lanosterol, C_1 transfer does not occur until after the three nuclear methyl groups are removed. The first alkylated sterol is 24-methylene-Δ^8-cholestenol (fecosterol), but in the genera *Candida* and *Agaricus* 24-methylenelanosterol appears as the first alkylated intermediate representing alkylation as the first step after cyclization. (For an expanded, detailed, and referenced consideration of this phenomenon, see Nes and McKean, 1977.) Such differences in sequencing may be of use phylogenetically. *Saccharomyces*, for instance, probably represents an evolutionary line which diverged from *Candida* and *Agaricus* earlier or later than *Candida* and *Agaricus* diverged from each other.

So far we have assumed that all reactions in two different sequences are the same and that the difference between the two depends on the point at which a given reaction occurs. A variant of this occurs when a new type of reaction is introduced in addition to or in place of one or more of the others. A well-examined case of this has to do with the cycloartenol–lanosterol bifurcation. Since the two pathways both give the same end products, and since the end products have the tetracyclic structure of lanosterol rather than that of the pentacyclic intermediate (cycloartenol), there has to be a corrective reaction (9,19-cyclo to Δ^8) in the cycloartenol case. However, there is no intrinsic reason why the correction should occur, say, at the beginning or toward the end of the sequence. As implied in the previous paragraph, this corrective action actually takes place at different points in the pathway in different plants. In euphorbs it seems to occur immediately after cyclization. In other plants it seems to be delayed until after both C_1 transfer and removal of one or both methyl groups at C-4 has occurred. The result of this is that we have two basic sequences represented simply by $A \rightarrow B \rightarrow C \rightarrow D$ and $A \rightarrow X \rightarrow Y \rightarrow Z$. A might be epoxysqualene, and B and X, respectively, lanosterol and cycloartenol. If D is the desired product, say, sitosterol, the A-to-Z sequence requires a crossover reaction that will convert X to B, Y to C, etc. The point at which this crossover from one sequence to the other occurs can be organismically and hence phylogenetically dependent. The case in which all the reactions are the same has similar though not identical qualities. Thus, the reduction of the Δ^{24} bond can occur at the beginning, in the middle, or at the end of the sequence, but here one is always dealing with the same reactions. In the simplest sense, one starts with the same product of cyclization, say, lanosterol, and has a 24-dehydro series up to the last reduction step giving cholesterol; or if reduction occurs immediately after lanosterol, one has a dihydro series that culminates directly in cholesterol. The former series is convertible at any point to the latter series by reduction. This sort of sequence problem is simpler than, and can be superimposed on, the one described with the cycloartenol–lanosterol bifurcation. In photosynthetic plants, both types of crossover between sequences can occur, and this leads to great subtlety in an analysis of sequence and its phylogenetic relationships. Much interesting work lies ahead of us in this area. The number of possible sequences between epoxysqualene and Δ^5 sterols is 10^2 to 10^3, which leaves ample opportunity to ask whether a given sequence is preferred in this or that organism, whether it is absolute, or whether all sequences occur, and finally to examine the evolutionary reasons that govern the choice.

B. FATTY ACID BIOSYNTHESIS

1. Orienting Remarks

Fatty acid biosynthesis comprises a series of reactions. (1) Reductive polymerization of acetate (via malonate) may proceed without interruption to a given stage on the synthetase or (2) may be interrupted to be followed by "chain elongation." (3) Aerobic, single or multiple desaturation or on rare cases (some bacteria) modification of the details of reductive polymerization arises to permit anaerobic desaturation. (4) Alkylation of a double bond in the manner which occurs in the steroidal side chain can occur, except that C_1 transfer appears only once in the fatty acid series, leading to introduction of a methylene, methyl, or cyclopropyl group. In addition, (5) the first carboxylic acid thioester ("initiator") to attack the enolate position of malonate may either be acetate leading to an even-numbered "backbone" or some other ester. When the next higher homolog (propionate) is the initiator, an odd-numbered chain is obtained. While this has not been studied extensively, the use of branched initiators derived from branched amino acids has been well investigated. The latter type of polymerization leads to the formation of odd- or even-numbered chains with a branch at one end (iso and anteiso acids). The extent to which living systems operate these four parts of fatty acid biosynthesis is variable and therefore offers an opportunity to examine phylogenetics.

2. Nonphotosynthetic Eukaryotes

The mammalian route is strongly limited with respect to the extent of polymerization and desaturation. Successive, nonstop condensation of the C_2 unit occurs only to the C_{18} stage, and only four double bonds can be subsequently introduced (Δ^4, Δ^5, Δ^6, and Δ^9). The specificities of the desaturases have been well studied (James, 1976). Two enzymes appear to exist for introduction of the Δ^9 bond. One promotes maximal conversion with a C_{18} chain and the other with a C_{14} chain (Gurr *et al.*, 1972). Both enzymes count from the carboxyl end of the chain so that regardless of the chain length desaturation occurs between the ninth and tenth carbon atoms, where C-1 is the carboxyl carbon. The double bond does not undergo methylation, as it does in certain other organisms, and methyl groups between C-5 and C-15 prevent introduction of a Δ^9 bond, indicating a high degree of selectivity. The other desaturases also count from the carboxyl group and seem to have rigid requirements. For instance, introduction of the Δ^9 bond must precede desaturation at C-6. Although C_2 polymerization only occurs as far as C_{18} in the saturated series, once a double bond is introduced chain elongation can occur. This permits the formation of longer fatty acids and shifts the double bonds toward the methyl end of the molecule, which is not possible by direct desaturation. Thus, C_{18}-Δ^9 plus C_2 gives C_{20}-Δ^{11}. The fatty acids, especially in the C_{18} and smaller series, derived by direct, *de novo* biosynthesis are used for membranous and other nonhormonal purposes. Elongation and further desaturation of ingested fatty acids is used for hormonal purposes (prostaglandins) (James, 1976; Samuelsson *et al.*, 1975). In the developing rat, for instance, a $\Delta^{5,8,11,14}$ fatty acid with a chain length of C_{19}, C_{20}, or C_{21} is needed for prostaglandin formation. The specificities of the desaturases and the necessity to add or subtract C_2 prevent the synthesis of any of these from an even-numbered acid with one or more of the allowed Δ^4, Δ^5, Δ^6, or Δ^9 bonds. This confers on the animal a dietary requirement for an acid ("essential fatty acid") that can enter the permitted reactions. Entry can be attained, for instance, with the C_{18}-$\Delta^{9,12}$ acid (linoleic acid), because it is allowed to proceed by Δ^6 desaturation to the $\Delta^{6,9,12}$ triene (γ-linolenic acid), then by chain elongation to the C_{20}-$\Delta^{8,11,14}$ acid and by desaturation at C-5 to the required C_{20}-$\Delta^{5,8,11,14}$ precursor (arachidonic acid) to the hormone. This result cannot be achieved with either oleic (Δ^9) or α-linolenic ($\Delta^{9,12,15}$) acids (Klenk, 1965; James, 1976). Since the rat cannot introduce a Δ^9 bond in the presence of Δ^{12}, the C_{18}-Δ^{12} acid is also inactive but the rat is unusual. Other mammals studied can convert Δ^{12} acids to the $\Delta^{9,12}$ analogs (but not Δ^9 to $\Delta^{9,12}$), and the former in the C_{17}, C_{18}, and C_{19} series actually do serve as essential fatty acids. In summary, mammals operate the fatty acid pathway so as to produce hormones and nonhormones. The pathway to the latter is direct (from C_2) and complete, while the pathway to the former lacks the initial phases. The animals either lost or never had the enzymes for the early part of the requisite pathway and depend on an exogenous source of precursors. Since the detail of utilization of dietary fatty acids for hormone production is not exactly the same for all mammalian species, further study of this subject in the vertebrate line more generally should prove worthwhile.

Insects have not been studied nearly as well as mammals. However, it is well established in several cases that they can make their nonhormonal fatty acids *de novo*. With the Dipterous *Ceratitis capitata*, for instance, triglycerides and phospholipids are rapidly labelled by $[1\text{-}^{14}C]$ acetate, and palmitic acid is aerobically desaturated to palmitoleic acid ($C_{16}\text{-}\Delta^9$) and incorporated into phospholipids (Municio *et al.*, 1971). With other insects the conversion of acetate to $C_{12:0}$, $C_{14:0}$, $C_{16:0}$, $C_{18:0}$, $C_{16:1}$, and $C_{18:1}$ and the elongation of $C_{16:0}$ and $C_{18:0}$ have been demonstrated, but the introduction of a Δ^{12} bond does not seem to occur at least in some cases. In a number of instances no linoleic acid ($\Delta^{9,12}$) or α-linolenic ($\Delta^{9,\,12,\,15}$) could be shown to be formed from acetate, and many insects are known to have a dietary requirement for polyunsaturates among which is α-linolenic acid. On the other hand, label from acetate has been reported to appear in small amounts in the linoleic acid fraction of some insects. It is not yet clear whether experimental problems are the cause of this discrepancy or whether it has to do with developmental or species differences. In summary, the present evidence suggests the existence of a basic similarity with mammals in the operation of the pathway as far as oleate for nonhormonal purposes, and a fundamental correlation also seems to exist in a requirement for exogenous unsaturates used presumably for hormonal and other specialized purposes.

The ciliated protozoan, *Tetrahymena pyriformis*, is quite unlike animals except that it carries acetate all the way to stearic and oleic acids (Koroly and Conner, 1976, and references cited). In addition, however, it converts oleate to linoleate and to γ-linolenate, representing the pathway Δ^0 to Δ^9 to $\Delta^{9,12}$ to $\Delta^{6,9,12}$; and the organism possesses a second pathway, in which palmitate ($C_{16:0}$) proceeds through desaturation (Δ^9) to palmitoleate and thence by chain elongation to *cis*-vaccenic acid ($C_{18}\text{-}\Delta^{11}$), which is further dehydrogenated to $C_{18:2}\text{-}\Delta^{6,11}$ as a major terminal product. Alternatively, chain elongation can occur to give $C_{20}\text{-}\Delta^{13}$. A third pathway also exists, in which various initiators are involved in place of acetate in the polymerization, leading to a complex fatty acid pattern composed in decreasing order of even-numbered *n*-alkanoic acids, odd iso acids, odd *n* acids, and odd anteiso acids. In summary, *T. pyriformis* has no requirement for exogenous fatty acid and biosynthesizes *de novo* a large array of such compounds. The pathway has two main bifurcations. One is at the very beginning and depends on the choice of initiator for polymerization. In the main route the choice is acetate, leading to the even-numbered series. It is in this series that the second bifurcation occurs. At the $C_{16:0}$ stage either desaturation at C-9 or chain elongation by two carbon atoms can occur. This is tantamount to a sequence problem. If desaturation at C-9 occurs first, the resulting $C_{16:1}$-Δ^9 yields the unusual $C_{18:2}$-$\Delta^{6,11}$ after chain elongation to C_{20}-Δ^{11} and dehydrogenation at C-6. If, on the other hand, chain elongation occurs first, the resulting $C_{18:0}$ yields $C_{18:3}$-$\Delta^{6,9,12}$ after desaturation at C-9, C-12, and C-6. The introduction of Δ^{12} in the latter case is intrinsically prohibited in the former case by the presence of the Δ^{11} bond, so we only have an ordering of two reactions (Δ^9 introduction and C_2 addition) involved. Whether the operation of both sequences is due to lack of specificity of the enzymes or to a positive control is not yet known but would be of considerable interest from an evolutionary point of view. The simplest explanation is the former, and if, as suspected on other grounds, the protozoa have had a separate history from as long ago as the Precambrian and therefore in a sense constitute a "living fossil" of primitive eukaryotes, a lack of specificity in the protozoan enzymes relative to mammals would correlate with a view in which evolution occurs with increasing enzymic specificities. In mammals $C_{16:1}$-Δ^9 does not seem to undergo chain elongation, consequentially unlike the protozoan case. However, enzymic specificities in terms of Michaelis constants clearly cannot be the whole difference between the protozoan and mammalian conditions, because the former has, and the latter lacks, a Δ^{12} desaturase. Here, we are dealing with the manner in which the enzyme counts from C-1, which almost certainly is determined by the positioning of the desaturating apparatus relative to the binding groups for the carboxylate thioester group. The ability of protozoa to position the desaturating apparatus in more places than can mammals is paralleled in the sterol field. In *T. pyriformis*, metabolism of cholesterol yields 22-dehydrocholesterol, which is unknown in vertebrates. Other differences between *T. pyriformis* and animals include the fact that, while both *T. pyriformis* and insects can dealkylate 24-ethylsterols, the pathways are different (Nes *et al.*, 1971), and *Tetrahymena* carries out anaerobic cyclization of squalene and fails to give lanosterol. *Tetrahymena* is, in fact, so different from the animal line that Nes (1974) has questioned whether it should be called "protozoa" in the sense of being a "first animal," and Holz (1966), in analyzing whether perhaps *Tetrahymena* is a plant, came to the conclusion after considering various of its biochemical characteristics "that *Tetrahymena* is *Tetrahymena*; it is neither fish nor fowl—nor geranium."

Based on fatty acid compositions, Korn *et al.* (1965) find that the amoeboid *Acanthamoeba*, *Hartmannella*, and *Physarium* (a "true" slime grown on fatty-acid-deficient media) have similar synthetic abilities. The $C_{18:1}$-Δ^9 acid arises *de novo* and then has much the same kind of sequence choices as described for *T. pyriformis*. Chain elongation leads to $C_{20:1}$-Δ^{11}, or desaturation at C-12 gives C_{18}-$\Delta^{9,12}$. The latter then is thought to proceed by chain elongation to $C_{20:2}$-$\Delta^{11,14}$ and on to arachidonic acid. Introduction of a Δ^{15} bond at the C_{18}-$\Delta^{9,12}$ stage fails to occur. While the detail is different, there are many similarities with *T. pyriformis*. In particular, a complete pathway exists to the necessary fatty acids. On the other hand, trypanosomes, e.g., *Trypanosoma lewisi*, which infect mammals, seem to be quite different. Labelled acetate leads only to $C_{20:5}$, $C_{22:5}$, and $C_{22:6}$ (presumably *via* $C_{18:0}$, which demonstrably proceeds to $C_{18:1}$-Δ^9, $C_{18:2}$-$\Delta^{9,12}$ and $C_{18:3}$-$\Delta^{6,9,12}$), despite dominance of (unlabelled) $C_{16:0}$, $C_{18:0}$, $C_{18:1}$-Δ^9, and C_{18}-$\Delta^{9,12}$. The dominant acids presumably are derived from the host (rat or blood). Growth does not occur on a medium devoid of fatty acid. Digenetic trypanosomes also absorb cholesterol from their animal hosts (see Nes and McKean, 1977, for a deeper discussion), which seems to inhibit endogenous production of ergosterol. Monogenetic trypanosomes, *e.g.*, *Crithidia*, are strikingly different from *Trypanosoma*. *Crithidia* synthesizes large, steady state amounts of $C_{18:0}$, $C_{18:1}$-Δ^9, $C_{18:2}$-$\Delta^{9,12}$, and $C_{18:3}$-$\Delta^{6,9,12}$ as well as $C_{22:5}$-$\Delta^{4,7,10,13,16}$ and small amounts of saturated acids that are probably branched. *Crithidia* is thought also to have a complete sterol pathway (to ergosterol).

The "cellular" slime molds in the class Acrasia, *e.g.*, *Dictyostelium discoideum* and *Polysphondylium pallidum,* are extremely unusual in that they biosynthesize $C_{16:2}$-$\Delta^{5,9}$, $C_{18:2}$-$\Delta^{5,9}$, and $C_{18:2}$-$\Delta^{5,11}$ (Korn *et al.*, 1965). These acids are the only polyunsaturates and account for 30% of the total fatty acids. Nothing is known about the pathways.

Fungi as a group have an extensive capacity to biosynthesize fatty acids *de novo* (Pyrrell, 1967; Wassef, 1977). Palmitic and stearic acids are the usual products of direct reductive polymerization of acetate. Chain elongation can then proceed. This has been demonstrated, for instance, with palmitic and arachidic acids in yeasts (*Saccharomyces* and *Candida*). Aerobic desaturation occurs, as demonstrated in *Saccharomyces, Neurospora,* and *Penicillium,* to give Δ^9 acids, e.g., oleate, with the enzyme counting from C-1. Oleate is then further desaturated via linoleate to α-linolenate or γ-linolenate. Depending on the organism, etc., these accumulate or proceed either aerobically to $C_{18:4}$ or successively by chain elongation and desaturation to give, ultimately, $C_{20:4}$-$\Delta^{5,8,11,14}$ (arachidonic acid) or $C_{22:6}$-$\Delta^{4,7,10,13,16,19}$ (docosahexenoic acid). The lower (classes Chytridomycetes and Oomycetes) and higher (class Zygomycetes) Phycomycetes both carry out polymerization, chain elongation, and desaturation, but the degree to which these processes occur varies. The lower ones tend to chain elongate more. Ascomycetes, Deuteromycetes, and Basidiomycetes resemble the higher Phycomycetes in tending not to chain elongate the C_{18} acids. In true yeasts one finds either few or no long-chain polyenoic acids at all. The Phycomycetes also seem to produce $C_{18:3}$-$\Delta^{6,9,12}$, while the other three kinds of fungi yield the $\Delta^{9,12,15}$ isomer if a triene is formed.

It appears that as the evolutionary ladder is climbed in the fungal world, chain elongation becomes restricted and the pathway of desaturation shifts toward the carboxyl end of the chain, but more detailed work on this subject, especially relating biosynthetic sequence to organism, is necessary.

3. Photosynthetic Eukaryotes

Photosynthetic eukaryotes, from algae through bryophytes and gymnosperms to angiosperms, all have the capacity to biosynthesize polyunsaturated acids *de novo* (Moreno *et al.*, 1971; Cherif *et al.*, 1975; Johns *et al.*, 1979; Kannangara and Stumpf, 1972). The pathway to α-linolenate ($\Delta^{9,12,15}$) from oleate (Δ^9) through linoleate ($\Delta^{9,12}$) appears to be a common property, but more than a single sequence may be involved. Evidence exists for one enzymatic system leading only to linoleate (CN^--inhibited) and another (CN^--insensitive) yielding linolenate. Also there are detailed differences with age, temperature, light, and plant parts. Roots, for instance, seem not to carry the pathway beyond the linoleate stage, and the major fatty acids of chloroplasts, linoleic and linolenic acids, are experimentally obtainable from acetate only with immature leaves. This misled earlier workers to believe that higher plants lacked appropriate biosynthetic capacity. Furthermore, *Chlorella* and the leaves of higher plants (but not photosynthetic bacteria) contain an unusual fatty acid (*trans*-Δ^3-C_{16}) believed to be localized in the chloroplast in the phosphatidyl glycerol fraction. Seeds also frequently show lipid biosynthesis that is not typical. Thus, Oo and Stumpf (1979) find the endosperm of the coconut (*Cocos nucifera* fruit) will not chain elongate C_8–C_{18} acids nor desaturate stearic acid. Moreover, coconut oil is more than 50% C_{12}–C_{14} acids, unlike most plant tissues, where the C_{16}–C_{18} acids are dominant. Conversely, Ihara and Tanaka (1978) find the oil of the pits of *Celtis sinensis* is 80% linoleic acid. Among the algae, the presence of 16:4 (ω3) and 18:1-Δ^{11} acids seems to be diagnostic for the Chlorophyta (Johns *et al.*, 1979).

The origin of arachidonic acid ($C_{20:4}$-$\Delta^{5,8,11,14}$) in the algae *Ochromonas danica* (Chrysophyceae), *Prophyridium cruentum* (Rhodophyceae), and *Euglena gracilis* (Euglenaceae) is an example of sequence that varies with the organismic type without altering the final structure (Nichols and Appleby, 1969). In the first two algae $C_{18:1}$-Δ^9 is first desaturated successively to $\Delta^{9,12}$ and $\Delta^{6,9,12}$ and then chain elongated to $C_{20:3}$-$\Delta^{8,11,14}$. Final introduction of the Δ^5 bond gives the end product. Chain elongation occurs one step earlier in *Euglena,* leading to the sequence $C_{18:1}$-Δ^9 to $C_{18:2}$-$\Delta^{9,12}$ to $C_{20:2}$-$\Delta^{11,14}$ to $C_{20:3}$-$\Delta^{8,11,14}$ to $C_{20:4}$-$\Delta^{5,8,11,14}$. The product of the second reaction in the sequence is different from that in the first case as a result of an inversion of reactions, but more than simple enzymatic specificity is involved, since the desaturation in question is at Δ^6 in the first case and Δ^8 in the second. This means a substantial change in the structure of the desaturase, as discussed earlier. It is not possible yet to explain the reasons for the difference in the ordering of the reactions. While it would be attractive to associate it with function, the relationship at least to photosynthesis is confused. In both *O. danica* and *E. gracilis* arachidonic acid is not associated with the photosynthetic apparatus (not in galactosyl diglyceride fraction and light not required), yet the pathways are different in the two organisms. Conversely, in *P. cruentum* (a red alga), *Monodus subterraneus* (Xanthophyceae), and mosses, arachidonic acid is associated with the chloroplast, and at least in the red algal case the pathway is the same (through $C_{18:3}$-$\Delta^{6,9,12}$) as in *O. danica*. The latter pathway also appears to operate in the nonphotosynthetic protomonad, *Crithidia*, based on the presence of $C_{18:3}$-$\Delta^{6,9,12}$ and no $C_{20:2}$-$\Delta^{11,14}$, but the other pathway is found in the nonphotosynthetic *Physarium* and *Acanthamoeba*, $C_{20:2}$-$\Delta^{11,14}$ but no $C_{18:3}$-$\Delta^{6,9,12}$ being present.

The nonphotosynthetic, parasitic angiosperms *Monotropa uniflora* and *M. hypopitys,* which in their sterol composition are main-line plants despite the lack of chloroplasts and operate the sterol pathway via cycloartenol, contain and presumably biosynthesize the usual fatty acids ($C_{16:0}$, $C_{16:1}$, $C_{18:0}$, $C_{18:1}$, $C_{18:2}$, and $C_{18:3}$) (Stanley and Patterson, 1977). The partially photosynthetic (early in life cycle) but parasitic *Cuscuta campestris* has similar fatty acids (and sterols), except that the proportion of $C_{18:3}$ is several times greater, and the nonphotosynthetic *Epifagus virginiana* makes $C_{18:3}$-$\Delta^{9,12,15}$ as do photosynthetic angiosperms (Nes *et al.*, 1979a).

4. Prokaryotes

Nichols *et al.* (1965) found that the blue-green algae *Anacystis nidulans* and *Anabena variabilis* both desaturate (Δ^9) and chain-elongate (by C_2) palmitic acid yielding palmitoleic, stearic, and oleic acids. *A. variabilis* was also shown, like its eukaryotic relative *Chlorella vulgaris*, to give labelled linoleic and α-linolenic acids from radioactive oleic acid and the latter from labeled stearic acid. No degradation of the acids occurred to, e.g., palmitic acid, in contrast to such degradation with photosynthetic bacteria. The latter ogranisms lack the ability to desaturate long-chain fatty acids directly, and degrade specifically labeled saturated acids to, presumably, acetate and resynthesize randomly labelled unsaturated acids (Harris *et al.*, 1965). The polar lipids of the blue-green algae (except for the absence of phosphatidyl ethanolamine, choline, and inositol, which are present in *Chlorella*) were found to be the same as in the eukaryotic algae (monogalactosyl diglyceride, phosphatidyl glycerol, digalactosyl diglyceride, and sulfoquinovosyl diglyceride). The latter four are common in chloroplasts. Other aspects of lipid biosynthesis in prokaryotes that can be inferred from occurrence are to be found in several earlier parts of this book. A great deal of variation is observed among bacteria. For instance, *Mycobacterium smegmatis* and *M. bovis* contain a series of unusual monounsaturated acids in the C_{14}–C_{26} series. The double bond is Δ^{10} (C_{14} and C_{16}), Δ^9 (C_{17}, C_{18}, and C_{19}), Δ^{11} (C_{20}), Δ^{13} (C_{22}), Δ^{15} (C_{24}), or Δ^{17} (C_{26}) (Hung and Walker, 1970). These structures suggest Δ^{10} desaturation of C_{14} and C_{16}, and Δ^9 desaturation of C_{17}, C_{18}, and C_{19}, the second of which is then successively elongated by C_2 units to C_{26}.

5. Unusual Fatty Acids

Although the biosynthesis of uncommon fatty acids has not been well studied, in passing we should like to mention them because the acquisition of an unusual enzyme might be used as a phylogenetic marker. An example is the ability to alkylate a fatty acid double bond by C_1 transfer. So far as we know, this is not terribly common, but the ability to alkylate follows anything but a clear evolutionary line. The C_{18}-10-methyl- and C_{18}-11,12-cyclopropyl acids (tuberculostearic and lactobacillic acids, respectively) are produced by bacterial metabolism. However, one also finds C_{17}-, C_{18}-, and C_{19}-cyclopropyl acids in arthropoda (Myriapoda; Diplopoda; and *Graphidostreptus*, a millipede) as well as in the Kingdom Photosynthetica among the angiosperms in the order Malvales. Despite the lack of correlation in general, there is a correlation in detail. Within the order Malvales, cyclopropyl fatty acids arise in several families, notably Sterculiaceae, Malvalaceae, Bombaceae, and Tiliaceae. Another group of unusual acids are those which have conjugated unsaturation (no interruption by a CH_2 group). Conjugated fatty acids have been found more or less randomly in seed oils of 11 of the 330 families, representing 9 of the 48 orders of higher plant, ranging from the primitive Santalaceae to the Compositae, which are thought to have arisen more recently. The botanical groups in which these fatty acids have been found are (order: family) Santales (Olaceae and Santalaceae), Rosales (Rosaceae), Geraniales (Euphorbiaceae), Sapindales (Balasaminaceae, Coriariaceae), Cucurbitales (Cucurbitaceae), Myrtiflorae (Punicaceae), Tubiflorae (Bignoniaceae), Dipsacales (Valerianaceae), and Campanulatae (Compositae). All species of the same family do not produce conjugated acids; a given chemical class, e.g., dienic, trienic, or acetylenic acids, is frequently found in more than one family; and a given family can have more than one class. Except for detail, such as that plants of one genus tend to produce the same conjugated acid, evolutionary relationships are not apparent. For instance, *cis,trans,cis*-trienes are found in the

Cucurbitaceae, Punicaceae, and Bignoniaceae, for which very different evolutionary relationships have been suggested on morphological grounds (Hutchinson, 1969).

C. HYDROCARBON BIOSYNTHESIS

Two major pathways to hydrocarbons exist (Major and Blomquist, 1978). In higher plants and insects (carefully investigated in the cockroach, *Periplaneta*) polymerization, chain elongation, and decarboxylation occur successively, while in bacteria there is the suggestion that two fatty acid chains are condensed at the C-1 end with decarboxylation occurring in the course of the condensation. It is obvious that no clear phylogenetic conclusions can be drawn.

CHAPTER 9

Conclusions

Our most important conclusion is that there is a great deal of interplay between many forces in the determination of what lipids shall be present in a given cell at a given time. Function seems to be one of the more important of these, and there is reason to believe that the idea of function cannot be isolated from environmental considerations. The various forces that seem to be operating have been discussed in connection with specific details. The data do not yet permit quantitative statements about how these forces interact, so we restrict our specific conclusions to the following list.

1. There can be little doubt that the universe, the earth, and life are all coevolving.

2. The question of the conditions under which life began, and indeed whether or not it actually originated on this earth, remains unsolved.

3. With the exception of meteorites, where the data leave much to be desired, lipids are not found elsewhere (the moon, Mars, intergalactic space).

4. Both organismic and molecular fossils indicate that life existed 3 billion years ago, and no unusual molecules have been found. The isopentenoid and polyacetate pathways were present in the earliest forms of life, which were prokaryotic. While some investigators think the earliest fossils include blue-green algae, others are less convinced. Organisms of undoubted blue-green-algal character, however, are known from about a billion years ago.

5. The organismic fossil record indicates that change occurred in time, but this has not been documented with molecular fossils. The important question as to whether and when biochemistry changed in time remains to be adequately explored.

6. The major forms of life that existed in the past still exist.

7. A fundamental basis for the idea of species is difficult to arrive at in terms of lipid biochemistry. The lipids of various species in a single genus can be very different, yet there does seem to be a degree of homology if one views the subject statistically and if a family of enzymes rather than a single

enzyme—or a single lipid—are considered. The problem we find in associating a coherent lipid pattern within a genus is paralleled by recent conclusions of Levin (1979), who examined the idea of species from a completely different point of view (population biology, seed dispersal, etc.). He concludes as a botanist that species, at least in the plant kingdom, "lack reality, cohesion, independence, and simple evolutionary or ecological roles." Diversity in lipid patterns among species of the same genus seems to be greater the less cellularly differentiated the organism is, but diversity is found even in as highly evolved forms as insects.

8. An attempt to discern commonality of lipid characters among organisms belonging to higher taxonomic groups, e.g., genera within a family, also has many difficulties.

9. No lipid characters absolutely associate all prokaryotes with each other or allow one to discriminate prokaryotes as a group from eukaryotes. This fails to correlate with recent information on the structures of RNA (Darnell, 1978; Hori and Osawa, 1979; Fox *et al.*, 1977) that shows prokaryotes, including blue-green algae, to be very similar to each other and distinct from eukaryotes. All eukaryotes, for instance, have a type of 5S RNA containing a loop (a U type of structure) with a ring of about a dozen nucleotides at the bottom. This ring is absent in all prokaryotic 5S RNA. Similarly, the structures and other aspects of mRNA in eukaryotes and prokaryotes is quite different, so much so as to lead Darnell (1978) to suggest that "sequential prokaryotic to eukaryotic evolution is unlikely." Strangely compared to this is that in terms of fatty acids and sterols the blue-green algae closely mimic eukaryotic algae and are very different from photosynthetic bacteria, which would lead to the conclusion that the two types of prokaryotes had different evolutionary histories and that the prokaryotic and eukaryotic algae had related histories. It is not yet clear whether the lipid or RNA data is the better parameter. If the RNA data is taken, we have to conclude that parallel and independent evolution of lipids occurred. It would seem clear, though, that the acquisition of certain lipid characters, say, the ability to form a polyunsaturated fatty acid or a sterol, occurred independently of the acquisition of certain ribonucleotide characters. Nevertheless, in some cases there does seem to be an association, since the existence of a certain kind of 16S ribosomal RNA appears to occur only in bacteria with isopentenoidal ethers in their membranes instead of fatty acid esters.

10. We are unable to see how the assumption of "random mutation with survival of the fittest" is helpful in understanding the complexities of lipid biosynthesis, structure, and distribution among organisms. Instead, the data suggest to us the operation of a number of principles interacting with each other in a goal-directed manner.

11. The data suggest that the dual criteria of the extent of completion of a

pathway, and the kinetic control of it such that intermediates are kept to small, steady-state concentrations, may be helpful in assessing the evolutionary status of an organism. Those forms of life that do not complete the pathway or that accumulate intermediates or both may be more primitive. Statistically, for instance, the unicellular blue-green algae do not complete the fatty acid pathway to polyunsaturates, whereas the colonial and filamentous species do; and in the embryophyte line, accumulation of steroid intermediates is common in the mosses and rare in tracheophytes. Since eukaryotic algae do not accumulate steroid intermediates, there is the suggestion that evolution has begun anew at the beginning of the embryophyte line.

12. While all examined organisms can polymerize acetate to fatty acids, the sterol pathway is far from ubiquitous. It is present in all photosynthetic organisms except photosynthetic bacteria, but it is absent in many nonphotosynthetic forms of life. In the sense of the previous point (11) its absence is a primitive character, yet many animals lack the pathway. In fact, it is found fully only in molluscs, echinoderms, vertebrates, and those fungi which are not oomycetous. A case can be made for the introduction of the pathway into the living milieu only after about a billion years ago, but it is not without its problems.

13. The pathway to sterols via cycloartenol or lanosterol divides the living world, respectively, into the evolutionary kingdoms Photosynthetica and Nonphotosynthetica without regard to other parameters. Since we are dealing with an intermediate here and not a functional end product, the bifurcation in the pathway serves as one of our best phylogenetic markers. Nonphotosynthetic flowering plants, for instance, are, as their flowers would suggest, derived from the photosynthetic line, since the cycloartenol route is used. Furthermore, the use of the cycloartenol route in nonphotosynthetic tissue of a functionally photosynthetic plant weighs heavily against the symbiotic origin of higher plants in fungi and blue-green algae. The reason for the existence of the two pathways remains to be discovered, but one possibility is that it is a time dependence. Early acquisition may have dictated the cycloartenol route and late acquisition the lanosterol route.

14. There is a correlation between the presence of a nervous system (animals) and the absence of certain biosynthetic steps. In particular, animals that biosynthesize sterols do not alkylate the Δ^{24} bond, and animals lacking biosynthesis either have a preference for dietary sterols lacking alkylation or they have a mechanism for removing the alkyl group. Plants, whether photosynthetic or not, have the capacity to make both alkylated and unalkylated sterols. Evolution of both pathways (reduction vs. alkylation of Δ^{24}) in plants was presumably possible due to the lack of strictures imposed by a nervous system. The animal line is also characterized by additional restriction. Generally, the number of different kinds of lipids biosynthesized

or absorbed from the diet is small. This is especially true in vertebrates and reaches extreme levels in mammals and man, where both the fatty acid and isopentenoid pathways fail to give the enormous array of lipids found in other forms of life. Among the many examples are the operation of the sterol pathway, in all animals studied, only to cholesterol; the inability of mammals to desaturate fatty acids at more than four positions; and the absence in all animals of squalene and epoxysqualene cyclases that give pentacycles. The nervous system seems to have transcended the need for the many different lipids required by plants. A more detailed analysis of the lipids of lower animals and a comparison with the ecological niches that they occupy should be instructive in relation to the way the animal line became restricted.

15. There appear to be two major reasons for the occurrence of lipids: genealogy and function. Each of these is documented in special cases, but the subject needs more attention before we can perceive how the two interrelate. This is not likely to be a simple task. The division of labor in the world seems to be determined by factors beyond our present understanding. No photosynthetic organism has a nervous system, for instance, even though for organisms of an appropriate surface area a nervous system would seem to be a highly desirable characteristic. At a molecular level, despite the beauty of our perception of the lipid bilayer in membranes, we now find bacteria with lipid monolayers. The evidence suggests that what is critical is the 40-Å dimension; how the dimension is achieved becomes a detail and is determined by other factors. Similarly, the tetracyclic sterol can be replaced in protozoan membranes by a pentacycle. Here again the flat, approximately 20-Å dimension of the molecule seems to be what is important rather than whether a sterol or pentacyclic triterpenoid is involved. A final example is the lipid chain extending from chlorophyll. This can be a fatty alcohol, farnesol, geranylgeraniol, or phytol. In all cases the dimensions are roughly the same, i.e., a chain of about 18 C atoms. Function, thus, involves major considerations without which there will be no function at all, and minor considerations as how best to fine tune what is happening in relation to the detail of the function both within the organism and in relation to other organisms.

16. The available evidence strongly suggests a careful mating of the biosynthesis of lipids with the function they are to play, and there seems to be a hierarchy in structure. Based on the statistics of biosynthesis and dietary requirements, sterols seem to be better membranous components than pentacyclic triterpenoids; and yeast will not accept tetrahymanol in place of sterol, while *T. pyriformis* will permit sterol to replace tetrahymanol. Sophistication in the mating of structure and function seems to increase with classic evolutionary status in some cases. For instance, discrimination against various exogenous sterols appears to be less in invertebrates than in vertebrates.

17. The chirality at C-24 of sterols has evolutionary correlates. Substituents at this position are more likely to have the β configuration in lower organisms than in higher ones. However, chirality in terpenes is associated with the size of the molecule rather than the type of organism it occurs in. The larger the molecule, the more likely is it to find one and the same enantiomeric series in all organisms carrying out biosynthesis.

18. The same compound can be derived by more than one route, and the route has organismic correlates. For instance, the manner in which the 24β-methyl group of ergosterol arises in *Chlorella* is different from the process operating in *Saccharomyces*. Phylogenetic relationships, unfortunately, are not as clear as this example would suggest, since some algae set the β configuration at C-24 in the same manner as the fungi do.

19. Although there are indeed strong distinctions between great groups of organisms in their biochemistry, acquisition of a particular enzyme or set of enzymes does not seem to have absolute familial qualities. The biosynthesis of triiodothyronine, which would generally be taken to be a characteristic of vertebrates, occurs in red algae. A prostaglandin has been found in onions. Conversely, the terpenes associated classically with pines and other higher plants are biosynthesized by insects.

References

Adams, D. F., Hessel, S. J., Judy, P. F., Stein, J. A., and Abrams, H. L. (1976). *Science* **192,** 465.

Adler, J. H., Young, M., and Nes, W. R. (1977). *Lipids* **12,** 364.

Adler, R. D., Metzger, A. L., and Grundy, S. M. (1974). *Gastroenterology* **66,** 1212.

Alais, J., Lablache-Combier, A., Lacoste, L., and Vidal, G. (1976). *Phytochemistry* **15,** 49.

Albrecht, P., and Ourisson, G. (1969). *Science* **163,** 1192.

Albrecht, P., and Ourisson, G. (1971). *Angew. Chem.* **10,** 209.

Albro, P. W., and Dittmer, J. C. (1969). *Biochemistry* **8,** 394.

Albro, P. W., and Dittmer, J. C. (1970). *Lipids* **5,** 320.

Alcaide, A., Viala, J., Pinte, F., Itoh, M., Nomura, T., and Barbier, M. (1971). *C. R. Acad. Sci. Paris Ser. C* **273,** 1386.

Allais, J. P., and Barbier, M. (1971). *Experentia* **27,** 506.

Allen, C. F., Good, P., and Holton, R. W. (1970). *Plant Physiol.* **46,** 748.

Amdur, R. H., Szabo, E. I., and Socransky, S. S. (1978). *J. Bacteriol.* **135,** 161.

Anders, E., Hayatsu, R., and Studier, M. (1973). *Science* **182,** 781.

Andersen, N. H., Ohta, Y., Liu, C-B., Kramer, C. M., Allison, K., and Huneck, S. (1977a). *Phytochemistry* **16,** 1727.

Andersen, N. H., Bissonette, P., Liu, C-B., Shunk, B., Ohta, Y., Tseng, C-L. W., Moore, A., and Huneck, S. (1977b). *Phytochemistry* **16,** 1731.

Anderson, I. G., Haslewood, G. A. D., Oldham, R. S., Amos, B., and Tokes, L. (1974). *Biochem. J.* **141,** 485.

Anderson, M. M. (1972). *Can. J. Earth Sci.* **9,** 1710.

Anderson, R., Kates, M., and Volcani, B. E. (1979). *Biochim. Biophys. Acta* **573,** 557.

Anding, C., Brandt, R. D., and Ourisson, G. (1971). *Eur. J. Biochem.* **24,** 259.

Anding, C., Parks, L. W., and Ourisson, G. (1974). *Eur. J. Biochem.* **43,** 459.

Andreasen, A. A., and Stier, T. J. B. (1953). *J. Cell Comp. Physiol.* **41,** 23.

Andreasen, A. A., and Stier, T. J. B. (1954). *J. Cell Comp. Physiol.* **43,** 271.

Arntzen, C. J., and Briantais, J. (1975). In *Bioenergetics of Photosynthesis* (Govindjee, ed.), Academic Press, New York.

Arthur, H. R., and Ko, P. D. S. (1974). *Phytochemistry* **13,** 2551.

Asselineau, C. P., Montrozier, H. L., Prome. J. C., and Savagnac, A. M. (1972). *Eur. J. Biochem.* **28,** 102.

Attaway, D., and Parker, P. L. (1970). *Science* **169,** 674.

Bada, J. L., Schroeder, R. A., and Carter, G. F. (1974). *Science* **184,** 791.

Baisted, D. J., Capstack, E., and Nes, W. R. (1962). *Biochemistry* **1,** 537.

Baker, G. L., Vroman, H. E., and Padmore, J. B. (1963). *Biochem. Biophys. Res. Commun.* **13**, 360.

Bandaranayake, W. M., Karunanayake, S., Sotheeswaran, S., Sultanbawaa, M. U. S., and Balasubraniam, S. (1977). *Phytochemistry* **16**, 699.

Barghoorn, E. S., and Schopf, J. W. (1965). *Science* **150**, 337.

Barghoorn, E. S., Troughton, J. H., and Margulis, L. (1974). *Am. Sci.* **62**, 389.

Barghoorn, E. S., and Tyler, S. A. (1965). *Science* **147**, 563.

Bennett, C. L., Beukens, R. P., Glover, M. R., Elmore, D., Gove, H. E., Kilius, L., Litherland, A. E., and Purser, P. H. (1978). *Science* **201**, 345.

Benveniste, P., Hirth, L., and Ourisson, G. (1965). *Bull. Soc. Fr. Physiol. Veg.* **11**, 252.

Benveniste, P., Hirth, L., and Ourisson, G. (1966). *Phytochemistry* **5**, 45.

Berkner, L. V., and Marshall, L. C. (1965). *Proc. Nat. Acad. Sci. U.S.A.* **53**, 1215.

Bernhard, K., von Bulow-Kostner, J., and Wagner, H. (1959). *Helv. Chim. Acta* **42**, 152.

Bertojo, M., Chui, M. F., Townes, C. H. (1974). *Science* **184**, 619.

Bird, C. W., Lynch, J. M., Pirt, F. L., Reid, W. W., Brooks, C. J. W., and Middleditch, B. (1971). *Nature* **230**, 473.

Bisalputra, T. (1974). In *Algal Physiology and Biochemistry* (W. D. P. Stewart, ed.), University of California Press, Berkeley.

Bitz, M. C., and Nagy, B. (1966). *Proc. Nat. Acad. Sci. U.S.A.* **56**, 1383.

Bjerve, K. S., and Bremer, J. (1969). *Biochim. Biophys. Acta* **176**, 570.

Bjorkhem, I., and Karlmar, K. E. (1974). *Biochim. Biophys. Acta* **337**, 129.

Bloch, K. (1965). *Science* **150**, 19.

Bloch, K. (1969). *Acc. of Chem. Res.* **2**, 193.

Bloch, K., and Vance, D. (1977). *Ann. Rev. Biochem.* **46**, 263.

Blomquist, G. J., Soliday, C. L., Byers, C. L., Braake, B. A., and Jackson, L. L. (1972). *Lipids* **7**, 356.

Bloomfield, D. K., and Bloch, K. (1960). *J. Biol. Chem.* **235**, 337.

Blumer, M. (1950). *Helv. Chim. Acta* **6**, 1627.

Blumer, M. (1965). *Science* **149**, 722.

Blumer, M., and Snyder, W. D. (1965). *Science* **150**, 1588.

Boiteau, P., Pasich, B., and Ratsimamanya, R. (1964). *Les Triterpenoides en Physiologie Vegetale et Animale,* Gauthiers-Villars, Paris.

Bold, H. C., and Wynne, M. J. (1978). *Introduction to the Algae,* Prentice-Hall, Englewood Cliffs, N.J.

Bolger, L. M., Rees, H. H., Ghisalberti, E. L., Goad, L. J., and Goodwin, T. W. (1970a). *Tetrahedron Lett.* 3043.

Bolger, L. M., Rees, H. H., Ghisalberti, E. L., Goad, L. J., and Goodwin, T. W. (1970b). *Biochem. J.* **118**, 197.

Bouvier, P., Rohmer, M., Benveniste, P., and Ourisson, G. (1976). *Biochem. J.* **159**, 267.

Bowers, W. S. (1978). *Lipids* **13**, 736.

Brady, R. O. (1968). *Adv. Clin. Chem.* **11**, 1.

Brenner, G. J. (1976). In *Origin and Early Evolution of Angiosperms* (C. B. Beck, ed.), p. 23. Columbia University Press, New York.

Briggs, M. H. (1963). *Nature* **197**, 1290.

Brock, T. D. (1973). *Science* **179**, 480.

Brock, T. D. (1974). In *Bergey's Manual of Determinative Bacteriology* (R. E. Buchanan and N. E. Gibbons, eds.), 8th ed., pp. 461–462. Williams and Wilkins, Baltimore.

Brock, T. D. (1978). *Thermophilic Microorganisms and Life at High Temperatures*, Springer-Verlag, New York.

Brown, H. (1949). *Revs. Mod. Phys.* **21**, 625.

Buchanan, R. E., and Gibbons, N. E. eds. (1974). *Bergey's Manual of Determinative Bacteriology* 8th ed., Williams and Wilkins, Baltimore.

Bu'Lock, J. D., and Osagie, A. U. (1976). *Phytochemistry* **15**, 1249.

Burlingame, A. L., and Simoneit, B. R. (1968). *Science* **160**, 531.

Burlingame, A. L., Calvin, M., Henderson, J. H. W., Reed, W., and Simoneit, B. R. (1970). *Science* **167**, 751.

Calvin, M. (1969a). *Chemical Evolution*, chap. 4–6. Clarendon Press, Oxford University Press, Oxford.

Calvin, M. (1969b). In *Perspectives in Biology and Medicine* **13**, 45.

Capstack, E., Baisted, D. J., Newschwander, W. W., Blondin, G. A., Rosin, N. L., and Nes, W. R. (1962). *Biochemistry* **1**, 1178.

Capstack, E., Rosin, N., Blondin, G. A., and Nes, W. R. (1965). *J. Biol. Chem.* **240**, 3258.

Cason, J., and Graham, D. W. (1965). *Tetrahedron* **21**, 471.

Castenholz, R. W. (1969). *J. Phycol.* **5**, 360.

Castle, M., Blondin, G. A., and Nes, W. R. (1963). *J. Am. Chem. Soc.* **85**, 3306.

Castle, M., Blondin, G. A., and Nes, W. R. (1967). *J. Biol. Chem.* **242**, 5796.

Catalano, S., Marsili, A., Morelli, I., and Pacchiani, M. (1976). *Phytochemistry* **15**, 1178.

Chandler, F. R., and Hooper, S. N. (1979). *Phytochemistry* **18**, 711.

Chardon-Loriaux, I., Morisaki, M., and Ikekawa, N. (1976). *Phytochemistry* **15**, 723.

Cheng, A. L. S., Kasperbauer, M. J., and Rice, L. G. (1971). *Phytochemistry* **10**, 1481.

Cherif, A., Dubacq, J. P., Mache, R., Oursel, A., and Tremolieres, A. (1975). *Phytochemistry* **14**, 703.

Christie, W. W. (1970). In *Topics in Lipid Chemistry* (F. D. Gunstone, ed.), Vol. I, pp. 1–49. Logos Press, London.

Chuecas, L., and Riley, J. P. (1969). *J. Mar. Biol. Ass. U.K.* **49**, 97.

Clayton, R. N. (1963). *Science* **140**, 192.

Cloud, P. E., Jr. (1965). *Science* **148**, 27.

Cloud, P. E., Jr. (1968). *Science* **160**, 729.

Cloud, P. E., Jr. (1974). *Am. Sci.* **62**, 389.

Cloud, P. E., Jr., Licari, G. R., Wright, L. A., and Troxel, B. W. (1969). *Proc. Nat. Acad. Sci. U.S.A.* **62**, 623.

Cloud, P. E., Jr., Wright, J., and Glover, L., III (1976). *Am. Sci.* **64**, 396.

Conner, R. L., Landrey, J. R., Joseph, J. M., and Nes, W. R. (1978). *Lipids* **13**, 692.

Conner, R. L., Mallory, F. B., Landrey, J. R., and Iyengar, C. W. L. (1969). *J. Biol. Chem.* **244**, 2325.

Cooper, W. J., and Blumer, M. (1968). *Deep-Sea Res.* **15**, 535.

Cordell, G. A. (1974). *Phytochemistry* **13**, 2343.

Crawford, A. R. (1974). *Science* **184**, 1179.

Crepet, W. L., Dilcher, D. L., and Potter, F. W. (1974). *Science* **185**, 781.

Cronin, J. R., and Moore, C. B. (1971). *Science* **172**, 1327.

Cronquist, A. (1968). *The Evolution and Classification of Flowering Plants*, Houghton-Mifflin, Boston.

Darland, G., and Brock, T. D. (1971). *J. Gen. Microbiol.* **67**, 9.

Darnell, J. E., Jr. (1979). *Science,* **202**, 1257.

Das, S. K., and Smith, E. D. (1968). *Ann. N.Y. Acad. Sci.* **147**, 411.

Davies, W. H., Mercer, E. I., and Goodwin, T. W. (1965). *Phytochemistry* **4**, 741.

Day, E. A., Malcorn, G. T., and Beeler, M. F. (1969). *Metab. Clin. Exp.* **18**, 646.

Devys, M., Alcaide, A., and Barbier, M. (1969). *Phytochemistry* **8**, 1441.

Dicke, R. H., Peebles, P. J. E., Roll, T. G., and Wilkinson, D. T. (1965) *Astrophys. J.* **142**, 414.

Dickerson, R. E., and Geis, I. (1969). *The Structure and Action of Proteins,* pp. 59–66. Harper and Row, New York.

Dietz, R. S., and Holden, J. C. (1970). *J. Geophys. Res.* **75**, 4939.

Dilcher, D. L., Pavlick, R. J., and Mitchell, J. (1970). *Science* **168**, 1447.

Doyle, J. A., and Hickey, L. J. (1976). In *Origin and Early Evolution of Angiosperms* (C. B. Beck, ed.), pp. 139–206. Columbia University Press, New York.

Doyle, P. J., Patterson, G. W., Dutky, S. R., and Thompson, M. J. (1972). *Phytochemistry* 11, 1951.

Dreyer, D. L. (1966). *Phytochemistry* 5, 367.

Dulaney, E. L., Stapley, E. O., and Simpf, K. (1954). *Appl. Microbiol.* 2, 371.

Eaton, G. P., Christiansen, R. L., Iyer, H. M., Pitt, A. M., Mabey, D. R., Blank, H. R., Zietz, I., and Gettings, M. E. (1975). *Science* 188, 787.

Eglinton, G., Douglas, A. G., Maxwell, J. R., Ramsay, J. N., and Stallberg-Stenhagen, S. (1966). *Science* 153, 1133.

Eglinton, G., and Murphy, M. (1969). *Organic Geochemistry*, Springer-Verlag, New York.

Ehrendorfer, F. (1971). In *Lehrbuch der Botanik für Hochschulen*, G. Fischer, Stuttgart.

Ehrhardt, J. D., Hirth, L., and Ourisson, G. (1967). *Phytochemistry* 6, 815.

Elliott, C. G. (1968). *J. Gen. Microbiol.* 51, 137.

Elliott, C. G. (1972). *J. Gen. Microbiol.* 72, 321.

Elliott, C. G., Hendrie, M. R., Knights, B. A., and Parker, W. (1964). *Nature (London)* 203, 427.

Elvoson, J., and Vagelos, P. R. (1969). *Proc. Nat. Acad. Sci. U.S.A.* 62, 957.

Emiliani, C., Hudson, J. H., Shinn, E. A., and George, R. Y. (1978). *Science* 202, 627.

Eppenberger, U., Hirth, L., and Ourisson, G. (1969). *Eur. J. Biochem.* 8, 180.

Erwin, J. A. (1973). *Lipids and Biomembranes of Eukaryotic Microorganisms*, Academic Press, New York.

Erwin, J. A., and Bloch, K. (1964). *Science* 143, 1006.

Fast, P. G. (1966). *Lipids* 1, 209.

Fenical, W. (1975). *J. Phycol.* 11, 245.

Ferezou, J. P., Devys, M., Allais, J. P., and Barbier, M. (1974). *Phytochemistry* 13, 593.

Fiertel, A., and Klein, H. P. (1959). *J. Bacteriol.* 78, 738.

Fieser, M., and Fieser, L. F. (1959). *Steroids*, Reinhold, New York.

Flanagan, V. P., and Ferretti, A. (1974). *Lipids* 9, 471.

Fogg, G. E., Stewart, W. D. P., Fay, P., and Walsby, A. E. (1973). *The Blue-Green Algae*, Academic Press, New York.

Folsome, C. E., Lawless, J. G., Romiez, M., and Ponnamperuma, C. (1973). *Geochim. Cosmochim. Acta* 37, 455.

Fox, G. E., Magrum, L. J., Balch, W. E., Wolfe, R. S., and Woese, C. R. (1977). *Proc. Nat. Acad. Sci. U.S.A.* 74, 4537.

Frasinel, C., Patterson, G. W., and Dutky, S. R. (1978). *Phytochemistry* 17, 1567.

Frayha, G. J. (1974). *Comp. Biochem. Physiol.* 49B, 93.

Fujino, Y., and Ohnishi, M. (1979). *Lipids* 14, 663.

Fujita, S. (1970). *Bull. Chem. Soc. Japan* 43, 2630.

Fujita, S. (1979). Abstract No. 56 of the Biological Chemistry Division, Annual Meeting of the American Chemical Society, Honolulu.

Fulco, A. J. (1970). *Biol. Chem.* 245, 2985.

Fulco, A. J. (1977). In *Polyunsaturated Fatty Acids* (W.-H. Kunau and R. T. Holman, eds.), The American Oil Chemists Society, Champagne, Ill.

Furuya, T., Nagumo, T., Itoh, T., and Kaneko, H. (1977). *Agric. Biol. Chem.* 41, 1607.

Galliard, T. (1973). In *Form and Function of Phospholipids* (G. B. Ansell, J. N. Hawthorne, and M. L. Dawson, eds.), pp. 261–266. American-Elsevier, New York.

Gammon, R. H. (1978). *Chem. Eng. News* 56, 21.

Gamow, G. (1945). *The Birth and Death of the Sun*, New American Library, New York.

Gamow, G. (1952). *The Creation of the Universe*, Viking, New York.

Geller, M. J. (1978). *Am. Sci.* 66, 176.

Gellerman, J. I., Anderson, W. H., and Schlenk, H. (1972). *Bryologist* **75**, 550.

Gelman, N. S., Lukoyanova, M. A., and Ostrouskii, D. N. (1975). *Biomembranes*, Vol. 6, Plenum Press, New York.

Gelpi, E., and Oro, J. (1970). *Geochim. Cosmochim. Acta* **34**, 981.

Gelpi, E., Schneider, H., Mann, J., and Oro, J. (1970a). *Phytochemistry* **9**, 603.

Gelpi, E., Han, J., Nooner, D. W., and Oro, J. (1970b). *Geochim. Cosmochim. Acta* **34**, 965.

Gershengorn, M. C., Smith, A. R. H., Goulston, G., Goad, L. J., Goodwin, T. W., and Haines, T. H. (1968). *Biochemistry* **7**, 1698.

Ghisalberti, E. L., de Souza, N. J., Rees, H. H., and Goodwin, T. W. (1970). *Phytochemistry* **9**, 1817.

Gibbons, G. F., Goad, L. J., Goodwin, T. W., and Nes, W. R. (1971). *J. Biol. Chem.* **246**, 3967.

Gloe, A., Pfennig, N., Brockmann, H., Jr., and Trowitzsch, W. (1975). *Arch. Microbiol.* **102**, 103.

Gloe, A., and Risch, N. (1978). *Arch. Microbiol.* **118**, 153.

Glover, L., III, and Sinha, A. K. (1973). *A. J. Sci.* **273-A**, 234.

Goad, L. J., and Goodwin, T. W. (1972). In *Progress in Phytochemistry* (L. Reinhold and Y. Liwschitz, eds.), Vol. 3, p. 113. Interscience, New York.

Goad, L. J., Knapp, F. F., Lenton, J. R., and Goodwin, T. W. (1974). *Lipids* **9**, 582.

Goodwin, T. W. (1961). Symposium No. III, Reports of the V International Congress of Biochemistry, Moscow, p. 1.

Goodwin, T. W. (1971). *Aspects of Terpenoid Chemistry and Biochemistry*, Academic Press, New York.

Gordon, G. S., Fitzpatrick, M. E., and Lubich, W. P. (1967). *Trans. Assoc. Am. Physicians* **80**, 183.

Grant, P. R., Grant, B. R., Smith, J. N. M., Abbott, I. J., and Abbott, L. K. (1976). *Proc. Nat. Acad. Sci. U.S.A.* **73**, 257.

Gray, J., Laufeld, S., and Boucot, A. J. (1974). *Science* **185**, 260.

Greig, J. B., Varma, K. R., and Caspi, E. (1971). *J. Am. Chem. Soc.* **93**, 760.

Gurr, M. I., and Brawn, P. (1970). *Eur. J. Biochem.* **17**, 19.

Gurr, M. I., Robinson, M. P., James, A. T., Morris, L. J., and Howling, D. (1972). *Biochim. Biophys. Acta* **280**, 415.

Haines, T. H. (1973). *Ann. Rev. Microbiol.* **27**, 403.

Haines, T. H. (1979). In *CRC Handbook of Microbiology* (A. I. Laskin and H. Lechevalier, eds.), 2nd ed., Vol. 5, CRC Press, West Palm Beach, Fla.

Haines, T. H., Pousada, M., Stern, B., and Mayers, G. L. (1969). *Biochem. J.* **113**, 565.

Hallam, A. (1975). *A Revolution in the Earth Sciences*, Clarendon Press, Oxford University Press, London.

Hamberg, M., Samuelsson, B., Bjorkhem, I., and Danielsson, H. (1974). In *Molecular Mechanisms of Oxygen Activation* (O. Hayaishi, ed.), p. 29, Academic Press, New York.

Hamilton, P. B. (1965). *Nature* **205**, 284.

Han, J., and Calvin, M. (1969). *Proc. Nat. Acad. Sci. U.S.A.* **64**, 436.

Han, J., Simoneit, B. R., Burlingame, A. L., and Calvin, M. (1969). *Nature (London)* **222**, 364.

Hansen, R. P. (1967). *Chem. Ind.* 39.

Hanson, J. R. (1972). In *Progress in Phytochemistry* (L. Reinhold and Y. Liwschitz, eds.), Vol. 3, p. 231. Interscience, New York.

Harborne, J. B. (1968). In *Progress in Phytochemistry* (L. Reinhold and Y. Liwschitz, eds.), Vol. 1, p. 545. Interscience, New York.

Harris, R. V., Wood, B. J. B., and James, A. T. (1965). *Biochem. J.* **94**, 22P.

Harvey, H. W. (1957). *Chemistry and Fertility of Sea Waters*, 2nd ed., Cambridge University Press, New York.

Haslewood, G. A. D. (1967). *Bile Salts*, Methuen, London.

Hayatsu, R. (1964). *Science* **146,** 1291.

Hayatsu, R. (1965). *Science* **149,** 443.

Hayatsu, R., Studier, M. H., and Anders, E. (1971). *Geochim. Cosmochim. Acta* **35,** 939.

Hayatsu, R., Studier, M. H., Matsuoka, S., and Anders, E. (1972). *Geochim. Cosmochim. Acta* **36,** 555.

Hayatsu, R., Studier, M. H., Moore, L. P., and Anders, E. (1975). *Geochim. Cosmochim. Acta* **39,** 471.

Hayes, J. M. (1967). *Geochim. Cosmochim. Acta* **31,** 1395.

Hayes, J. M., and Biemann, K. (1968). *Geochim. Cosmochim. Acta* **32,** 239.

Hitchcock, C. H. S., and Nichols, B. W. (1971). *Plant Lipid Biochemistry,* Academic Press, London.

Hodgson, G. W., and Baker, B. L. (1969). *Geochim. Cosmochim. Acta* **33,** 943.

Hoering, T. C., and Abelson, P. H. (1961). *Proc. Nat. Acad. Sci. U.S.A.* **47,** 623.

Holton, R. W., Blecker, H. H., and Stevens, T. S. (1968). *Science* **160,** 545.

Holz, G. G. (1966). *J. Protozol.* **13,** 2.

Holz, G. G., Jr., and Conner, R. L. (1973). In *Biology of Tetrahymena* (A. M. Elliott, ed.), p. 99. Dowden, Hutchinson, and Ross, Stroudsburg, Pa.

Hori, H., and Osawa, S. (1979). *Proc. Natl. Acad. Sci. U.S.A.* **76,** 381.

Horn, D. H. S. (1964). *Austr. J. Chem.* **17,** 464.

Horodyski, R. J., and Bloeser, B. (1978). *Science* **199,** 682.

Hubble, E. (1925). *Pop. Astron.* **33,** 252.

Hui, W.-H., and Li, M. M. (1976). *Phytochemistry* **15,** 427.

Hui, W.-H., and Li, M. M. (1977). *Phytochemistry* **16,** 111.

Hung, J. G. C., and Walker, R. W. (1970). *Lipids* **5,** 720.

Hunter, K., and Rose, A. H. (1972). *Biochim. Biophys. Acta* **260,** 639.

Hutchinson, J. (1969). *Evolution and Phylogeny of Flowering Plants,* Academic Press, New York.

Huxley, J. (1969). *The Wonderful World of Life,* pp. 1–96. Doubleday, Garden City, N.Y.

Ihara, S., and Tanaka, T. (1978). *J. Am. Oil Chem. Soc.* **55,** 471.

Ikan, R., and Kashman, J. (1963). *Israel J. Chem.* **1,** 502.

Ikan, R., and McLean, J. (1960). *J. Chem. Soc.,* 893.

Ikan, R., and Seckbach, J. (1972). *Phytochemistry* **11,** 1077.

Ikawa, S., and Tammar, A. R. (1976). *Biochem. J.* **153,** 343.

Ingram, D. S., Knights, B. A., McEvoy, T. J., and McKay, P. (1968). *Phytochemistry* **7,** 1241.

Itoh, T., Tamura, T., and Matsumoto, T. (1973). *J. Am. Oil Chem. Soc.* **50,** 122.

Itoh, T., Tamura, T., and Matsumoto, T. (1974). *Lipids* **9,** 173.

Itoh, T., Tamura, T., and Matsumoto, T. (1975). *Lipids* **10,** 454.

Ives, A. J., and O'Neill, A. N. (1958). *Can. J. Chem.* **36,** 434.

Jackson, L. L. (1970). *Lipids* **5,** 38.

Jackson, L. L. (1972). *Comp. Biochem. Physiol.* **41B,** 331.

Jackson, L. L., and Blomquist, G. J. (1976). In *Chemistry and Biochemistry of Natural Waxes* (P. E. Kolattukudy, ed.), Elsevier, Amsterdam.

Jain, M. K., and White, H. B. (1977). *Adv. Lipid Res.* **15,** 1.

James, A. T. (1976). In *Function and Biosynthesis of Lipids* (N. G. Bazan, R. R. Brenner, and N. M. Giusto, eds.), pp. 51–74. Plenum Press, New York.

Jamieson, G. R., and Reid, E. H. (1969). *Phytochemistry* **8,** 1489.

Jarolim, V., Heino, K., Hemmert, F., and Sŏrm, F. (1965). *Collect. Czech. Chem. Commun.* **30,** 873.

Jermy, A. C., Crabbe, J. A., and Thomas, B. A. (1973). *The Phylogeny and Classification of the Ferns,* Academic Press, New York.

Johns, R. B., Belsky, T., McCarthy, E. D., Burlingame, A. L., Huang, P., Schnoes, H. K., Richter, W., and Calvin, M. (1966). *Geochim. Cosmochim. Acta* **30,** 1191.

Johns, R. B., Nichols, P. D., and Perry, G. J. (1979). *Phytochemistry* **18**, 799.

Johnson, A. R., Pearson, J. A., Shenstone, F. S., and Fogerty, A. C. (1967). *Lipids* **2**, 308.

Kaneda, T. (1977). *Bacteriol. Rev.* **41**, 391.

Kaneko, H., Hosohara, M., Tanaka, M., and Itoh, T. (1976). *Lipids* **11**, 837.

Kannangara, C. G., and Stumpf, P. K. (1972). *Arch. Biochem. Biophys.* **148**, 414.

Kaplan, I. R., Degens, E. T., and Renten, J. H. (1963). *Geochim. Cosmochim. Acta* **27**, 805.

Karrer, W. (1958). *Konstitution und Vorkommen der Organischen Pflanzenstoffe*, Birkhauser Verlag, Berlin.

Kates, M. (1972). In *Ether Lipids: Chemistry and Biology* (F. Snyder, ed.), pp. 351–398. Academic Press, New York.

Kates, M., Park, E. E., Palameta, B., and Joo, C. N. (1971). *Can. J. Biochem.* **49**, 275.

Kellogg, T. F. (1975). *Comp. Biochem. Physiol.* **50B**, 109.

Kenyon, C. N (1972). *J. Bacteriol.* **109**, 827.

Kenyon, C. N., Ripka, R., and Stanier, R. Y. (1972). *Arch. Mikrobiol.* **83**, 216.

Kerr, R. A. (1978). *Science* **200**, 36.

Kinsella, J. E., Shimp, J. L., Mai, J., and Weinrauch, J. (1977). *J. Am. Oil Chem. Soc.* **54**, 424.

Klein, R. M., and Cronquist, A. (1967). *Q. Rev. Biol.* **42**, 105.

Kleinschmidt, M. G., and McMahon, V. A. (1970). *Plant Physiol.* **46**, 286.

Klenk, E. (1965). *Adv. Lipid Res.* **3**, 1.

Klenk, E. Knipprath, W., Eberhagen, D., and Koof, H. P. (1963). *Hoppe-Seyler's Z. Physiol. Chem.* **334**, 44.

Knights, B. A., and Berrie, A. M. M. (1971). *Phytochemistry* **10**, 131.

Knoche, H., and Ourisson, G. (1967). *Angew. Chem.* **79**, 1107.

Knoche, H., Albrecht, P., and Ourisson, G. (1968). *Angew. Chem.* **80**, 666.

Knoll, A. H., and Barghoorn, E. S. (1975). *Science* **190**, 52.

Korn, E. D., Greenblatt, C. L., and Lees, A. M. (1965). *J. Lipid Res.* **6**, 43.

Koroly, M. J., and Conner, R. L. (1976). *J. Biol. Chem.* **251**, 7588.

Kupchan, S. M., Meshylam, H., and Sneden, A. T. (1978). *Phytochemistry* **17**, 767.

Kvenvolden, K. A. (1967). *J. Am. Oil Chem. Soc.* **44**, 628.

Kvenvolden, K., Lawless, J., Pering, K., Peterson, E., Flores, J., Ponnamperuma, C., Kaplan, I. R., and Moore, C. (1970). *Nature* **228**, 923.

Kvenvolden, K., Lawless, J. and Ponnamperuma C. (1971). *Proc. Nat. Acad. Sci. U.S.A.* **68**, 486.

Lands, W. E. M., Hemler, M. E., and Crawford, C. G. (1977). In *Polyunsaturated Fatty Acids* (W-H. Kunau and R. T. Holman, eds.), The American Oil Chemists Society, Champagne, Ill.

Langworthy, T. A. (1977). *Biochim. Biophys. Acta* **487**, 37.

Langworthy, T. A., Smith, P. F., and Mayberry, W. R. (1972). *J. Bacteriol.* **112**, 1193.

Langworthy, T. A., Mayberry, W. R., and Smith, P. F. (1974). *J. Bacteriol.* **119**, 106.

Largeau, C., Goad, L. J., and Goodwin, T. W. (1977a). *Phytochemistry* **16**, 1931.

Largeau, C., Goad, L. J., and Goodwin, T. W. (1977b). *Phytochemistry* **16**, 1925.

Larson, R. B. (1977). *Am. Sci.* **65**, 188.

Lavie, D., and Glotter, E. (1971). *Fortschr. Chem. Org. Naturst.* **39**, 307.

Lawless, J. G., Kvenvolden, K. A., Peterson, E., Ponnamperuma, C., and Moore, C. (1971). *Science* **173**, 626.

Lemmon, R. M. (1970). *Chem. Rev.* **70**, 95.

Lenton, J. R., Hall, J., Smith, A. R. H., Ghisalberti, E. L., Rees, H. H., Goad, L. J., and Goodwin, T. W. (1971). *Arch. Biochem. Biophys.* **143**, 664.

Leventhal, J. S., and Threlkeld, C. N. (1978). *Science* **202**, 430.

Levin, D. A. (1979). *Science* **204**, 381.

Levin, E. Y., and Bloch, K. (1964). *Nature* **202**, 4927.

Levy, R. L., Grayson, M. A., and Wolf, C. J. (1973). *Geochim. Cosmochim. Acta* **37**, 467.

Liebecq, C. (1978). Chairman, IUB Editors, *Biochemical Nomenclature and Related Documents,* The Biochemical Society, London.

Lincoln, D. E., and Langenheim, J. H. (1976). *Biochem. Syst. Eco.* **4**, 237.

Lynen, F. (1967). *Biochem. J.* **102**, 381.

MacCarthy, J. J., and Patterson, G. W. (1974a). *Plant Physiol.* **54**, 129.

MacCarthy, J. J., and Patterson, G. W. (1974b). *Plant Physiol.* **54**, 133.

Magrum, L. J., Leuhrsen, K. W., and Woese, C. R. (1978). *Mol. Evol.* **11**, 1.

Mahadevan, V. (1978). *Prog. Chem. Fats Other Lipids* **15**, 255.

Majerus, P. W., and Vagelos, P. R. (1967). In *Advances in Lipid Research* (R. Paoletti and D. Kritchevsky, eds.), Vol. 5, p. 2. Academic Press, New York.

Major, M. A., and Blomquist, G. J. (1978). *Lipids* **13**, 323.

Malhotra, H. C., and Nes, W. R. (1971). *J. Biol. Chem.* **246**, 4934.

Malhotra, H. C., and Nes, W. R. (1972). *J. Biol. Chem.* **247**, 6243.

Mallory, F. B., Gordon, J. T., and Conner, R. L. (1963). *J. Am. Chem. Soc.* **85**, 1362.

Manzoor-I-Khuda, M. (1966). *Tetrahedron* **22**, 2377.

Masters, P. M., and Zimmerman, M. R. (1978). *Science* **201**, 811.

Maudinas, B., and Villoutreix, J. (1977). *Phytochemistry* **16**, 1299.

Maurice, A., and Baraud, J. (1967). *Rev. Fr. Corps Cras* **14**, 713.

Maxon, L. R., and Wilson, A. C. (1974). *Science* **185**, 66.

Mayberry-Carson, K. J., Langworthy, T. A., Mayberry, W. R., and Smith, P. F. (1974). *Biochim. Biophys. Acta* **360**, 217.

Mayr, E. (1978). *Sci. Am.* **239**, 47.

McCall, G. J. H. (1973). *Meteorites and Their Origins,* Wiley, New York.

McCarthy, E. D., and Calvin, M. (1967). *Nature* **216**, 642.

McCarthy, E. D., Van Hoeven, W., and Calvin, M. (1967). *Tetrahedron Lett.,* 4437.

McKean, M. L., and Nes, W. R. (1977a). *Phytochemistry* **16**, 683.

McKean, M. L., and Nes, W. R. (1977b). *Lipids* **12**, 382.

McMorris, T. C. (1978). *Lipids* **13**, 716.

Meinschein, W. G. (1965). *Science* **150**, 601.

Meinschein, W. G., Frondel, C., Laur, P., and Mislow, K. (1966). *Science* **154**, 377.

Meinschein, W. G., Cordes, E., and Shiner, W. J. (1970). *Science* **167**, 753.

Mellor, D. P. (1972). In *Chemistry in Space Research* (R. F. Landel and A. Rembaum, eds.), pp. 83–103. American Elsevier, New York.

Mercer, E. I., and Davies, C. L. (1974). *Phytochemistry* **13**, 1607.

Mercer, E. I., and Davies, C. L. (1975). *Phytochemistry* **14**, 1545.

Mercer, E. I., and Davies, C. L. (1979). *Phytochemistry,* in press.

Mercer, E. I., and Johnson, M. W. (1969). *Phytochemistry* **8**, 2329.

Metcalf, R. L. (1979). *Entomol. Soc. Am. Bull.* **25**, 30.

Mikolajczak, K. L. (1977). *Prog. Chem. Fats Other Lipids* **15**, 97.

Miller, R. W., Earle, F. R., and Wolff, I. A. (1965). *J. Am. Oil Chem. Soc.* **42**, 817.

Miller, S. L. (1953). *Science* **117**, 528.

Miller, S. L., and Urey, H. C. (1959). *Science* **130**, 245.

Mirov, N. T., Zavarin, E., and Snajberk, K. (1966). *Phytochemistry* **5**, 97.

Molnar, P., and Tapponnier, P. (1975). *Science* **189**, 419.

Morell, P., and Braun, P. (1972). *J. Lipid. Res.* **13**, 293.

Moreno, V. J., de Moreno, J. E. A., and Brenner, R. (1979). *Lipids* **14**, 15.

Mosettig, E., and Nes, W. R. (1955). *J. Org. Chem.* **20**, 884.

Mulheirn, L. J. (1973). *Tetrahedron Lett.,* 3175.

Muller, R. A. (1977). *Science* **196**, 489.

Muller, R. A., Stephenson, E. J., and Mast, T. S. (1978). *Science* **201**, 347.

Municio, A. M., Odriozola, J. M., Pineiro, A., and Ribera, A. (1971). *Biochim. Biophys. Acta* **248**, 212.

Murphy, M. T. J., McCormick, A., and Eglinton, G. (1967). *Science* **157**, 1040.

Murphy, R. C., Preti, C., Nafissi, M. M., and Biemann, K. (1970). *Science* **167**, 755.

Murray, J., and Thomson, A. (1977). *Phytochemistry* **16**, 465.

Nagy, B., and Bitz, C. M. (1963). *Arch. Biochem. Biophys.* **101**, 240.

Nagy, L. A. (1974). *Science* **183**, 514.

Nes, W. D., Patterson, G. W., Southall, M. A., and Stanley, J. L. (1979a). *Lipids* **14**, 274.

Nes, W. D., Patterson, G. W., and Bean, G. A. (1979b). *Lipids* **14**, 458.

Nes, W. R. (1971). *Lipids* **6**, 219.

Nes, W. R. (1974). *Lipids* **9**, 596.

Nes, W. R. (1977). In *Advances in Lipid Research* (R. Paoletti and D. Kritchevsky, eds.), Vol. 15, p. 233. Academic Press, New York.

Nes, W. R., and Ford, D. L. (1963). *J. Am. Chem. Soc.* **85**, 2137.

Nes, W. R., and McKean, M. L. (1977). *Biochemistry of Steroids and Other Isopentenoids,* University Park Press, Baltimore.

Nes, W. R., and Mosettig, E. (1954). *J. Am. Chem. Soc.* **76**, 3182.

Nes, W. R., Baisted, D. J., Capstack, E., Newschwander, W. W., and Russell, P. T. (1967). In *Biochemistry of the Chloroplast* (T. W. Goodwin, ed.), Vol. 2, p. 273. Academic Press, New York.

Nes, W. R., Malya, P. A. G., Mallory, F. B., Ferguson, K. A., Landrey, J. R., and Conner, R. L. (1971). *J. Biol. Chem.* **246**, 561.

Nes, W. R., Thampi, N. S., and Lin, J. T. (1972). *Cancer Res.* **32**, 1264.

Nes, W. R., Cannon, J. W., Thampi, N. S., and Malya, P. A. G. (1973). *J. Biol. Chem.* **248**, 484.

Nes, W. R., Alcaide, A., Mallory, F. B., Landrey, J. R., and Conner, R. L. (1975a). *Lipids* **10**, 140.

Nes, W. R., Krevitz, K., Behzadan, S., Patterson, G. W., Landrey, J. R., and Conner, R. L. (1975b). *Biochem. Biophys. Res. Commun.* **66**, 1462.

Nes, W. R., Adler, J. H., Sekula, B. C., and Krevitz, K. (1976a). *Biochem. Biophys. Res. Commun.* **71**, 1296.

Nes, W. R., Krevitz, K., and Behzadan, S. (1976b). *Lipids* **11**, 118.

Nes, W. R., Krevitz, K., Joseph. J., Nes, W. D., Harris, B., Gibbons, G. F., and Patterson, G. W. (1977a). *Lipids* **12**, 511.

Nes, W. R., Varkey, T. E., and Krevitz, K. (1977b). *J. Am. Chem. Soc.* **99**, 260.

Nes, W. R., Sekula, B. C., Nes, W. D., and Adler, J. H. (1978a). *J. Biol. Chem.* **253**, 6218.

Nes, W. R., Joseph J. M., Landrey, J. R., and Conner, R. L. (1978b). *J. Biol. Chem.* **253**, 2361.

Nicholas, H. J. (1973). In *Phytochemistry* (L. P. Miller, ed.), Vol. 2, p. 254. Van Nostrand Reinhold, New York.

Nichols, B. W. (1965). *Biochim. Biophys. Acta* **106**, 274.

Nichols, B. W., and Appleby, R. S. (1969). *Phytochemistry* **8**, 1907.

Nichols, B. W., Harris, R. V., and James, A. T. (1965). *Biochem. Biophys. Res. Commun.* **20**, 256.

Nicolaides, N. (1967). *Lipids* **2**, 266.

Nielsen, J. K., Larsen, L. M., and Sorensen, H. (1977). *Phytochemistry* **16**, 1519.

Nielsen, P. E., Nishimura, H., Liang, Y., and Calvin, M. (1979). *Phytochemistry* **18**, 103.

Nooner, D. W., and Oro, J. (1967). *Geochim. Cosmochim. Acta* **31**, 1359.

Oo, K. C., and Stumpf, P. K. (1979). *Lipids* **14**, 132.

Oparin, A. I. (1924). *Proiskhozhdenie zhizni,* Izd. Moskovskii Rabochii, Moscow.

Oparin, A. I. (1953). *Origin of Life,* 2nd ed., Dover, New York.

Oparin, A. I. (1957). *The Origin of Life on the Earth,* Academic Press, New York.

Oro, J. (1960). *Biochem. Biophys. Res. Commun.* **2**, 407.

Oro, J. (1961). *Nature* **190**, 389.

Oro, J., Gelpi, E., and Nooner, D. W. (1968). *J. Br. Interplanet. Soc.* **21**, 83.

Oro, J., Nooner, D. W., Zlatkis, A., Wikström, S. A., and Barghoorn, E. S. (1965). *Science* **148**, 77.

Oro, J., Tornabene, T. G., Nooner, D. W., and Gelpi, E. (1967). *J. Bacteriol.* **93**, 1811.

Oro, J., Nakaparksin, S., and Gil-Av, E. (1971). *Nature* **230**, 107.

Oshima, M., and Yamakawa, T. (1974). *Biochemistry* **13**, 1140.

Oudejans, R. C. H. M. (1972a). *J. Insect Physiol.* **18**, 857.

Oudejans, R. C. H. M. (1972b). *Comp. Biochem. Physiol.* **42B**, 15.

Oudejans, R. C. H. M., Van Der Horst, D. J., and Dongen, J. P. C. M. (1971a). *Biochemistry* **10**, 4938.

Oudejans, R. C. H. M., Van Der Horst, D. J., and Zandee, D. I. (1971b). *Comp. Biochem. Physiol.* **40B**, 1.

Oyama, V. I., Carle, G. C., Woeller, F., and Pollack, J. B. (1979). *Science* **203**, 803.

Pace-Asciak, C. (1977). *Prostaglandins* **13**, 811.

Palmer, M. A., Goad, L. J., Goodwin, T. W., Copsey, D. P., and Boar, R. B. (1978). *Phytochemistry* **17**, 1577.

Pant, P., and Rastogi, R. T. (1979). *Phytochemistry* **18**, 1095.

Paoletti, R., Galli, G., Grossi-Paoletti, E., Fiecchi, A., and Scala, A. (1971). *Lipids* **6**, 134.

Paoletti, C., Pushparaj, B., Florenzano, G., Capella, P., and Lercker, G. (1976a). *Lipids* **11**, 258.

Paoletti, C., Pushparaj, P., Florenzano, G., Capella, P., and Lercker, G. (1976b). *Lipids* **11**, 266.

Papenfuss, G. F. (1946). *Bull. Torrey Bot. Club* **73**, 217.

Parker, P. L. (1969). In *Organic Geochemistry* (G. Eglinton and M. Murphy, eds.), pp. 363–366. Springer-Verlag, New York.

Patt, T. E., and Hanson, R. S. (1978). *J. Bacteriol.* **134**, 636.

Patterson, G. W. (1967). *J. Phycol.* **3**, 22.

Patterson, G. W. (1969). *Comp. Biochem. Physiol.* **3**, 391.

Patterson, G. W. (1970). *Lipids* **5**, 597.

Patterson, G. W. (1971). *Lipids* **6**, 120.

Patterson, G. W. (1974). *Comp. Biochem. Physiol.* **47B**, 453.

Patterson, G. W. (1979). Reports of the International Symposium on Marine Algae of the Indian Ocean Region, Bhavnagar, India.

Patterson, G. W., and Krauss, R. W. (1965). *Plant Cell Physiol.* **6**, 211.

Patterson, G. W., Thompson, M. J., and Dutky, S. R. (1974). *Phytochemistry* **13**, 191.

Penzias, A. A. (1978). *Am. Sci.* **66**, 291.

Penzias, A. A., and Wilson, R. W. (1965). *Astrophys. J.* **142**, 419.

Pereira, W. E., Summons, R. E., Rindfleisch, T. C., Duffield, A. M., Zeitman, B., and Lawless, J. G. (1975). *Geochim. Cosmochim. Acta* **39**, 163.

Pering, K. L., and Ponnamperuma, C. (1971). *Science* **173**, 237.

Philippi, G. T. (1965). *Geochim. Cosmochim. Acta* **29**, 1021.

Phillips, T. L., Peppers, R. A., Avcin, M. A., and Laughnan, P. F. (1974). *Science* **184**, 1367.

Pickett-Heaps, J. D., and Marchant, H. J. (1972). *Cytobios* **6**, 255.

Playford, P. E., and Cockbain, A. E. (1969). *Science* **165**, 1008.

Pollock, G. E., Chang, C-N., Cronin, S. E., Kvenvolden, K. A. (1975). *Geochim. Cosmochim. Acta* **39**, 1571.

Ponnamperuma, C., Kvenvolden, K., Chang, S., Johnson, R., Pollock, G., Philpott, D., Kaplan, I., Smith, J., Schopf, J. W., Gehrke, C., Hodgson, G., Breger, J. A., Halpern, B., Duffield, A., Krauskopf, K., Barghoorn, E., Holland, H., and Keil, K. (1970). *Science* **167**, 760.

Ponsinet, G., and Ourisson, G. (1968). *Phytochemistry* **7**, 757.

Ponsinet, G., Ourisson, G., and Oehlschlager, A. C. (1968). In *Recent Advances in Phytochemistry* (T. J. Mabry, R. E. Alston, and V. C. Runeckler, eds.), Vol. 1, p. 271. Plenum Press, New York.

Prescott, G. W. (1968). *The Algae: A Review,* Houghton Mifflin, Boston.

Prost, M., Maume, B. F., and Padieu, P. (1974). *Biochim. Biophys. Acta* **360**, 230.

Pugh, E. L., and Kates, M. (1979). *Lipids* **14**, 159.

Pyrrell, D. (1967). *Can. J. Microbiol.* **13**, 755.

Raab, K. H., de Souza, N. J., and Nes, W. R. (1968). *Biochim. Biophys. Acta* **152**, 742.
Rahier, A., Cattel, L., and Benveniste, P. (1977). *Phytochemistry* **16**, 1187.
Rao, M. K., Perkins, E. G., Conner, W. E., and Bhattacharyya, A. K. (1975). *Lipids* **10**, 566.
Rattray, J. B. M., Schibeci, A., and Kidby, D. K. (1975). *Bacteriol. Rev.* **39**, 197.
Ray, P. H., White, D. C., and Brock, T. D. (1971a). *J. Bacteriol.* **108**, 227.
Ray, P. H., White, D. C., and Brock, T. D. (1971b). *J. Bacteriol.* **106**, 25.
Rees, H. H., Goad, L. J., and Goodwin, T. W. (1968). *Tetrahedron Lett.,* 273.
Rees, H. H., Goad, L. J., and Goodwin, T. W. (1969). *Biochim. Biophys. Acta* **176**, 892.
Rhoades, D. G., Lincoln, D. E., and Langenheim, J. H. (1976). *Biochem. Syst. Ecol.* **4**, 5.
Roberts, M. (1974). *Science* **183**, 371.
Rodriguez, E., and Levin, D. A. (1976). In *Recent Advances in Phytochemistry* (J. W. Wallace and R. L. Mansell, eds.), Vol. 10, p. 214.
Rohmer, M., and Brandt, R. D. (1973). *Eur. J. Biochem.* **36**, 446.
Rohmer, M., Ourisson, G., Benveniste, P., and Bimpson, T. (1975). *Phytochemistry* **14**, 727.
Rohmer, M., Bouvier, P., and Ourisson, G. (1979). *Proc. Nat. Acad. Sci. U.S.A.* **76**, 847.
de Rosa, M., Gambacorta, A., and Bu'Lock, J. D. (1974a). *J. Bacteriol.* **117**, 212.
de Rosa, M., Gambacorta, A., Minale, L., and Bu'Lock, J. D. (1971a). *Chem. Commun.* 619.
de Rosa, M., Gambacorta, A., Minale, L., and Bu'Lock J. D. (1971b). *Chem. Commun.* 1334.
de Rosa, M., Gambacorta, A., and Minale, L., and Bu'Lock, J. D. (1971c). *Chem. Commun.,* 619.
de Rosa, M., Minale, L., and Sodano, G. (1973). *Comp. Biochem. Physiol.* **45B**, 883.
de Rosa, M., Gambacorta, A., Minale, L., and Bu'Lock, J. D. (1974b). *Chem. Commun.* 543.
de Rosa, M., Gambacorta, A., and Bu'Lock, J. D. (1975). *Phytochemistry* **15**, 143.
de Rosa, M., Gambacorta, A., and Bu'Lock, J. D. (1976). *Phytochemistry* **15**, 143.
Rothstein, M. (1967). *Comp. Biochem. Physiol.* **21**, 109.
Roughan, P. G. (1975). *Lipids* **10**, 609.
Rubey, W. W. (1951). *Bull. Geol. Soc. Amer.* **62**, 111.
Rubinstein, I., and Goad, L. J. (1974). *Phytochemistry* **13**, 485.
Rudwick, M. J. S. (1972). *The Meaning of Fossils,* American Elsevier, New York.
Ruheman, S., and Raud, H. (1932). *Brennst. Chem.* **13**, 341.
Runnegar, B., and Pojeta, J., Jr. (1974). *Science* **186**, 311.
Russell, H. N. (1929). *Astrophys. J.* **70**, 11.
Russell, N. J., and Harwood, J. L. (1979). *Biochem. J.* **181**, 339.
Russell, P. T., van Aller, R. T., and Nes, W. R. (1967). *J. Biol. Chem.* **242**, 5802.
Sagan, C. (1961). *Organic Matter and the Moon,* N. A. S.-N. R. C. Publ. 757, Washington.
Salen, G., and Grundy, S. M. (1973). *J. Clin. Invest.* **52**, 2822.
Samuelsson, B., Granstrom, E., Green, K., Hambert, M., and Hannarstrom, S. (1975). *Ann. Rev. Biochem.* **44**, 669.
Sarjeant, W. A. S. (1974). *Fossil and Living Dinoflagellates,* Academic Press, New York.
Sawicki, E. H., and Pisano, M. A. (1977). *Lipids* **12**, 125.
Schopf, J. W. (1974). *Origins Life* **5**, 119.
Schopf, J. W. (1975). *Ann. Rev. Earth Planet. Sci.* **3**, 213.
Schopf, J. W., and Barghoorn, E. S. (1969). *J. Paleontol.* **43**, 111.
Schubert, K., Rose, G., and Horhold, C. (1967). *Biochim. Biophys. Acta* **137**, 168.
Schubert, K., Rose, G., Wachtel, H., Hörhold, C., and Ikekawa, N. (1968). *Eur. J. Biochem.* **5**, 246.
Schubert, L., Rose, G., Tummler, R., and Ikekawa, N. (1964). *Hoppe-Seyler's Z. Physiol. Chem.* **339**, 293.
Schwendinger, R. B. (1969). In *Organic Geochemistry* (G. Eglinton and M. Murphy, eds.), p. 428. Springer-Verlag, New York.
Schwendinger, R. B., and Erdman, J. G. (1964). *Science* **144**, 1575.

Schwenk, E., and Alexander, G. J. (1958). *Arch. Biochem. Biopys.* **76,** 65.

Seigler, D. S., and Butterfield, C. S. (1976). *Phytochemistry* **15,** 842.

Sever, J., and Parker, P. L. (1969). *Science* **164,** 1052.

Shaw, R. (1966). *Adv. Lipid Res.* **4,** 107.

Simpson, G. G. (1953). *The Major Features of Evolution,* Columbia University Press, New York.

Skrigan, A. I. (1951). *Dokl. Akad. Nauk. SSSR* **80,** 607.

Skrigan, A. I. (1964). *Tr. Vses. Nauchn. Tekhn. Soveshch. Gorki* **1963,** 108.

Simmonds, P. G., Bauman, A. J., Bollin, E. M., Gelpi, E., and Oro, J. (1969). *Proc. Nat. Acad. Sci. U.S.A.* **64,** 1027.

Smith, C. L., and Rand, C. S. (1975). *Science* **190,** 1105.

Smith, G. M. (1955). *Cryptogamic Botany,* Vol. 1, p. 546. McGraw-Hill, New York.

Smith, J. W., and Kaplan, I. R. (1970). *Science* **167,** 1367.

Smith, T. M., Brooks, T. J., and Lockard, V. G. (1970). *Lipids* **5,** 854.

Southall, M., Motta, J. J., and Patterson, G. W. (1977). *Am. J. Bot.* **64,** 246.

de Souza, N. J., and Nes, W. R. (1969). *Phytochemistry* **8,** 819.

Stanley, J. L., and Patterson, G. W. (1977). *Phytochemistry* **16,** 1611.

Stasek, C. R. (1972). *Chem. Zool.* **6,** 1.

Stewart, K. D., and Mattox, K. R. (1975). *Bot. Rev.* **41,** 104.

Stransky, K., Streibl, M., and Herout, V. (1967). *Collect. Czech. Chem. Commun.* **32,** 3213.

Streibl, M., and Herout, V. (1969). In *Organic Geochemistry* (G. Eglinton and M. Murphy, eds.), Chaps. 16 and 17. Springer-Verlag, New York.

Studier, M., Hayatsu, R., and Anders, E. (1965). *Science* **149,** 1455.

Studier, M., Hayatsu, R., and Anders, E. (1966). *Science* **152,** 106.

Studier, M. H., Hayatsu, R., and Anders, E. (1968). *Geochim. Cosmochim. Acta* **32,** 151.

Studier, M. H., Hayatsu, R., and Anders, E. (1972). *Geochim. Cosmochim. Acta* **36,** 189.

Stumpf, P. K. (1976). In *Plant Biochemistry* (J. Bonner and J. E. Varner, eds.), Academic Press, New York.

Stumpf, P. K., and Weber, N. (1977). *Lipids* **12,** 120.

Sucrow, W., and Girgensohn, B. (1970). *Chem. Ber.* **103,** 750.

Sucrow, W., and Reimerdes, A. (1968). *Z. Naturforsch.* **23b,** 42.

Sucrow, W., Schubert, B., and Richter, W. (1974). *Chem. Ber.* **104,** 3689.

Suzue, G., Tsukada, K., Nakai, C., and Tanaka, S. (1968). *Arch. Biochim. Biophys.* **123,** 644.

Svoboda, J. A., Pepper, J. H., and Baker, G. L. (1966). *J. Insect Physiol.* **12,** 1549.

Svoboda, J. A., Thompson, M. J., and Robbins, W. E. (1971). *Nature New Biol. (London)* **230,** 57.

Svoboda, J. A., Thompson, M. J., Robbins, W. E., and Kaplanis, J. N. (1978). *Lipids* **13,** 742.

Tansey, M. R., and Brock, T. D. (1972). *Proc. Nat. Acad. Sci., U.S.A.* **69,** 2426.

Taylor, M. E. (1966). *Science* **153,** 198.

Tchen, T. T., and Bloch, K. (1955). *J. Am. Chem. Soc.* **77,** 6085.

Teshima, S. I., and Kanazawa, A. (1974). *Comp. Biochem. Physiol.* **47B,** 555.

Tornabene, T. (1978). *J. Mol. Evol.* **11,** 253.

Tornabene, T. G., Bennet, E. O., and Oro, J. (1967). *J. Bacteriol.* **94,** 344.

Tornabene, T. G., and Langworthy, T. A. (1979). *Science* **203,** 51.

Tornabene, T. G., Wolfe, R. S., Balch, W. E., Holzer, G., Fox, G. E., and Oro, J. (1978). *J. Mol. Evol.* **11,** 259.

Trainor, F. R. (1978). *Introductory Phycology,* p. 525. Wiley, New York.

Tsai, L. B., and Patterson, G. W. (1976). *Phytochemistry* **15,** 1131.

Tsuda, Y., Morimoto, A., Sano, T., Inubushi, Y., Mallory, F. B., and Gordon, J. T. (1965). *Tetrahedron Lett.* **19,** 1427.

Turfitt, G. F. (1943). *Biochem. J.* **42,** 376.

Urey, H. C. (1952). *The Planets: Their Origin and Development,* Yale University Press, New Haven, Conn.

Urey, H. C. (1966). *Science,* **151,** 157.

Urey, H. C., and Lewis, J. S. (1966). *Science* **152,** 102.

Valentine, J. W., and Campbell, C. A. (1975). *Am. Sci.* **63,** 673.

Van Aller, R. T., Chikamatsu, H., de Souza, N. J., and Nes, W. R. (1968). *Biochem. Biophys. Res. Commun.* **31,** 842.

Van Aller, R. T., Chikamatsu, H., de Souza, N. J., and Nes, W. R. (1969). *J. Biol. Chem.* **244,** 6645.

Van Der Horst, D. J. (1970). *Neth. J. Zool.* **20,** 433.

Van Der Horst, D. J., and Oudejans, R. C. H. M. (1972). *Comp. Biochem. Physiol.* **41B,** 823.

Van Der Horst, D. J., Oudejans, R. C. H. M., and Zandee, D. I. (1972). *Comp. Biochem. Physiol.* **41B,** 417.

Van Der Horst, D. J., and Voogt, P. A. (1969). *Comp. Biochem. Physiol.* **31,** 763.

Van Dorsselaer, A., Ensminger, A., Spyckerelle, C., Dastillung, M., Sieskind, O., Arpino, P., Albrecht, P., Ourisson, G., Brooks, P. W., Gaskell, S. J., Kimble, B. J., Philip, R. P., Maxwell, J. R., and Eglington, G. (1974). *Tetrahedron Lett.* 1349.

Volpe, J. J., and Vagelos, P. R. (1973). *Ann. Rev. Biochem.* **42,** 21.

Voogt, P. A., Van De Ruit, J. M., and Van Rheenen, J. W. A. (1974). *Comp. Biochem. Physiol.* **48B,** 47.

Voogt, P. A., Van Rheenen, J. W. A., and Zandee, D. I. (1975). *Comp. Biochem. Physiol.* **B50,** 511.

Waldrop, M. (1979). *Chem. Eng. News,* Feb. 19 issue, 26.

Walker, J. W., and Skvaria, J. J. (1975). *Science* **187,** 445.

Walton, M. J., and Pennock, J. F. (1972). *Biochem. J.* **127,** 471.

Wassef, M. K. (1977). *Adv. Lipid Res.* **15,** 159.

Weete, J. D. (1974). In *Fungal Lipid Biochemistry, Monographs in Lipid Research* (D. Kritchevsky, ed.), p. 39. Plenum Press, New York.

Weete, J. D., and Laseter, J. L. (1974). *Lipids* **9,** 575.

Wegener, A. (1966). *The Origin of Continents and Oceans,* Dover, New York.

Weier, T. E., Harrison, A. H. P., and Risley, E. B. (1965). *J. Ultrastruct. Res.* **13,** 92.

Weier, T. E., Stocking, C. R., and Shumway, L. K. (1966). *Brookhaven Symp. Biol.* **19,** 353.

Weigelt, J., and Noack, K. (1931). *Nova Acta Leopoldina,* **1,** 87.

Wetherill, G. W. (1979). *Sci. Am.* **240,** 54.

Weyl, P. K. (1978). *Science* **202,** 475.

Wiik, H. B. (1956). *Geochim. Cosmochim. Acta* **9,** 279.

Willett, J. D., and Downey, W. L. (1974). *Biochem. J.* **138,** 233.

Wilton, D. C., Watkinson, I. A., and Akhtar, M. (1970). *Biochem. J.* **119,** 673.

Winters, K., Parker, P. L., and Baalen, C. V. (1969). *Science* **163,** 467.

Woese, C. R., and Fox, G. E. (1977). *Proc. Nat. Acad. Sci. U.S.A.* **74,** 5088.

Woese, C. R., Magrum, L. J., and Fox, G. E. (1978). *J. Mol. Evol.* **11,** 245.

Wolk, C. P. (1973). *Bacteriol. Rev.* **37,** 32.

Wood, B. J. (1974). In *Algal Physiology and Biochemistry* (W. D. P. Stewart, ed.), University of California Press, Berkeley.

Wood, B. J. B., Nichols, B. W., and James, A. T. (1965). *Biochim. Biophys. Acta* **106,** 261.

Wooton, J. A. M., and Wright, L. D. (1962). *Comp. Biochem. Physiol.* **5,** 253.

Yamamoto, S., Mitsuhashi, M., Endo, S., Abe, M., and Sugiyama, N. (1979). Abstract 17 of the Division of Biological Chemistry, Annual Meeting of the American Chemical Society, Honolulu.

Yokokawa, H. (1979). Abstract 57 of the Division of Biological Chemistry, Annual Meeting of the American Chemical Society, Honolulu.

Yokoyama, A., Natori, S., and Aoshima, K. (1975). *Phytochemistry* **14,** 487.

Zander, J. M., Greig, J. B., and Caspi, E. (1970). *J. Biol. Chem.* **245,** 1247.

Zavarin, E., and Snajberk, K. (1965). *Phytochemistry* **4,** 141.

Author Index

Subject Index